Introductory Fisheries Analyses with R

Chapman & Hall/CRC
The R Series

Series Editors

John M. Chambers
Department of Statistics
Stanford University
Stanford, California, USA

Torsten Hothorn
Division of Biostatistics
University of Zurich
Switzerland

Duncan Temple Lang
Department of Statistics
University of California, Davis
Davis, California, USA

Hadley Wickham
RStudio
Boston, Massachusetts, USA

Aims and Scope

This book series reflects the recent rapid growth in the development and application of R, the programming language and software environment for statistical computing and graphics. R is now widely used in academic research, education, and industry. It is constantly growing, with new versions of the core software released regularly and more than 6,000 packages available. It is difficult for the documentation to keep pace with the expansion of the software, and this vital book series provides a forum for the publication of books covering many aspects of the development and application of R.

The scope of the series is wide, covering three main threads:

- Applications of R to specific disciplines such as biology, epidemiology, genetics, engineering, finance, and the social sciences.
- Using R for the study of topics of statistical methodology, such as linear and mixed modeling, time series, Bayesian methods, and missing data.
- The development of R, including programming, building packages, and graphics.

The books will appeal to programmers and developers of R software, as well as applied statisticians and data analysts in many fields. The books will feature detailed worked examples and R code fully integrated into the text, ensuring their usefulness to researchers, practitioners and students.

Published Titles

Stated Preference Methods Using R, *Hideo Aizaki, Tomoaki Nakatani, and Kazuo Sato*

Using R for Numerical Analysis in Science and Engineering, *Victor A. Bloomfield*

Event History Analysis with R, *Göran Broström*

Computational Actuarial Science with R, *Arthur Charpentier*

Statistical Computing in C++ and R, *Randall L. Eubank and Ana Kupresanin*

Basics of Matrix Algebra for Statistics with R, *Nick Fieller*

Reproducible Research with R and RStudio, Second Edition, *Christopher Gandrud*

R and MATLAB® *David E. Hiebeler*

Nonparametric Statistical Methods Using R, *John Kloke and Joseph McKean*

Displaying Time Series, Spatial, and Space-Time Data with R, *Oscar Perpiñán Lamigueiro*

Programming Graphical User Interfaces with R, *Michael F. Lawrence and John Verzani*

Analyzing Sensory Data with R, *Sébastien Lê and Theirry Worch*

Parallel Computing for Data Science: With Examples in R, C++ and CUDA, *Norman Matloff*

Analyzing Baseball Data with R, *Max Marchi and Jim Albert*

Growth Curve Analysis and Visualization Using R, *Daniel Mirman*

R Graphics, Second Edition, *Paul Murrell*

Introductory Fisheries Analyses with R, *Derek H. Ogle*

Data Science in R: A Case Studies Approach to Computational Reasoning and Problem Solving, *Deborah Nolan and Duncan Temple Lang*

Multiple Factor Analysis by Example Using R, *Jérôme Pagès*

Customer and Business Analytics: Applied Data Mining for Business Decision Making Using R, *Daniel S. Putler and Robert E. Krider*

Implementing Reproducible Research, *Victoria Stodden, Friedrich Leisch, and Roger D. Peng*

Graphical Data Analysis with R, *Antony Unwin*

Using R for Introductory Statistics, Second Edition, *John Verzani*

Advanced R, *Hadley Wickham*

Dynamic Documents with R and knitr, Second Edition, *Yihui Xie*

Introductory Fisheries Analyses with R

Derek H. Ogle

Northland College

Ashland, Wisconsin, USA

CRC Press
Taylor & Francis Group
Boca Raton   London   New York

CRC Press is an imprint of the
Taylor & Francis Group, an **informa** business

A CHAPMAN & HALL BOOK

CRC Press
Taylor & Francis Group
6000 Broken Sound Parkway NW, Suite 300
Boca Raton, FL 33487-2742

© 2016 by Taylor & Francis Group, LLC
CRC Press is an imprint of Taylor & Francis Group, an Informa business

No claim to original U.S. Government works

Printed on acid-free paper
Version Date: 20151008

International Standard Book Number-13: 978-1-4822-3520-3 (Hardback)

Visit the Taylor & Francis Web site at
http://www.taylorandfrancis.com

and the CRC Press Web site at
http://www.crcpress.com

To my mentors – Nick, Jim, and George

Contents

Preface

In 2004, I discovered the R environment (R Development Core Team 2015b). After some self-instruction, I became interested in how to apply this remarkable tool to fisheries science. Within a year, I had briefly introduced R to a fisheries class at the University of Minnesota and to professionals at a Minnesota Chapter of the American Fisheries Society workshop. Emboldened by the reception at these different venues and convinced that R (or, really, any script-based language) was *the* way to analyze data, I began to develop code and examples to demonstrate analyses typical of an advanced undergraduate or early graduate fisheries population dynamics curriculum. That code evolved into the **FSA** package and those examples became the "analytical vignettes" on the *fishR* website (http://derekogle.com/fishR/). Later, I added worked examples in R to supplement the influential *Analysis and Interpretation of Freshwater Fisheries Data* book (Guy and Brown 2007).

Many people from around the world have either cited **FSA** or *fishR* or have let me know that these resources are useful for their purposes. Additionally, several instructors, and students in their classes, have used these resources to aid their teaching and learning. The use of **FSA** and *fishR* have exceeded my expectations and demonstrated either a need or a desire in the fisheries profession to use R for fisheries-related analyses.

With this momentum, it is now time to formalize the examples from *fishR* into a book that will reach either a different or a wider audience. I hope that this work will demonstrate the flexibility and power of R, provide insight into the reproducibility of script-based analyses, and help more fisheries scientists use R to be more efficient and productive.

Book Description and Intended Audience

This book is a "how-to" guide for conducting common fisheries-related analyses in R. For each method, enough background information is provided so that the reader is assured of the specific details of the implementation of the method; however, neither an in-depth description nor a derivation from first principles is provided. For example, the Petersen capture-recapture population estimator (Section 9.2) is not derived from the first principles of equality of ratios or from maximum likelihood estimation that originates from the theory of binomial sampling. However, the specific formulae and the assumptions of the method are described, along with a demonstration of how to perform the analysis in R. The reader is directed to other common resources for foundational material. The book deviates slightly from this philosophy in instances where a clear description is not available in other common resources (e.g., symmetry tests for comparing two age estimates in Section 4.3). Thus, the reader will generally need to enter this book knowing which analysis to perform, but wanting to know how to perform that analysis in R.

The analyses presented in this book are commonly performed by practicing fisheries scientists and are also present in many graduate and upper-level undergraduate fisheries science, analysis, or management courses. Most topics are covered at the introductory or intermediate level (similar to the levels of Guy and Brown (2007) and Haddon (2011), but below the level of Quinn II and Deriso (1999)) and, thus, should be attainable by advanced undergraduate students and all fisheries professionals. Some reviewers asked to include more advanced materials. Some of these requests (e.g., mixed models, Bayesian analyses, catch-at-age models) were not filled as I tried to contain the size of the book and keep it at a consistent introductory level.

There are two primary audiences for this book. First, fisheries professionals who largely know which analysis they want to perform, but want to do that analysis in R, will find this book to be a useful reference. Second, instructors (and their students) will find this book to be a practical supplement for specifying how to perform analyses in R that are described in their main text or demonstrated with a proprietary software that is unavailable.

The first three chapters provide a minimal introduction to the R environment that builds a foundation for the remaining chapters on fisheries-specific analyses. More detailed introductions to the R environment can be found in (among many) Dalgaard (2002), Verzani (2014), and the *Introduction to R* document that is distributed with the R environment. The reader should be familiar with the content of Chapters 1–3 before reading other chapters. Chapters 4 and 5 describe methods commonly used to provide data for topics in later chapters. The material in these two chapters is not a prerequisite for later chapters and, thus, they may be read independently. Linear regression, a foundational statistical method in many fisheries-specific analyses, is introduced within the context of the weight-length relationship in Chapter 7. The linear regression material in Chapter 7 should be mastered before Chapters 8, 10, and 11, which all utilize linear regression. Similarly, nonlinear regression is introduced within the context of analyzing individual growth in Chapter 12 and is then used to examine the stock-recruit relationship in Chapter 13.

Companion Website

A website to support the book is at `http://derekogle.com/IFAR/`. Data files and all of the code in the code boxes for each chapter are available there. The scripts contain some code that does not appear in, but was used to produce, the printed book. That code is commented as such in the scripts. One can, of course, download the scripts and run the code to reproduce the results shown in the book. However, a deeper understanding of the material will be gained if you interact with the code more closely — that is, type and run the code to create your own scripts with your own comments.

The companion website also contains supplemental code and text for topics that provide further examples of the material in the book. These materials, referred to in the book as the "online supplement," are likely of interest to the fisheries scientist but are beyond the introductory level of the book. Practice exercises for each chapter are also available on the website.

The companion website contains updated descriptions and code for functions within base R or any packages used in the book that may have changed since the book was printed. Finally, an errata, if one is needed, will also be maintained there.

Software Information and Conventions

The analyses in this book use the R environment (R Development Core Team 2015b). Many of the fisheries-specific analyses require functions from the **FSA** (Fisheries Stock Assessment) package, which I maintain. Instructions for installing **FSA** are on the companion website.

In the main text, names of R packages are in bold text (e.g., **FSA**); function names, function arguments, and code are in a courier font (e.g., `mrClosed()`); and external file names are in italicized courier font (*SMBassWB.txt*). In-line references to function names are followed by empty parentheses (e.g., `mrClosed()`) and arguments are followed by an equals sign (e.g., `method=`).

Code and resulting output are displayed in lightly shaded boxes (e.g., see below this paragraph). Input code is preceded by the usual R command prompt (i.e., `>`) with the results (i.e., output) shown immediately after the input on lines with no preceding characters. With these conventions, the input and output should look very similar to the results in your R console.

```
> sd(1:10)
[1] 3.02765
```

Most figures in the book use the following modifications of the default base graphing parameters (most of which are described in detail in Section 3.6.1):

```
> par(mar=c(3.05,3.05,0.65,0.65),mgp=c(1.9,0.3,0),tcl=-0.2,
     las=1,cex.lab=0.95,cex.axis=0.9)
```

This book was written in LaTeX (Lamport 1994) supported by **knitr** (Xie 2013). It was compiled under the R setup shown below. Other packages were added as needed in each chapter. Citations for R packages only appear at the first mention of the package.

```
R version 3.2.2 (2015-08-14)
Platform: i386-w64-mingw32/i386 (32-bit)
Running under: Windows 7 x64 (build 7601) Service Pack 1

attached base packages:
[1] stats      graphics  grDevices utils      datasets  methods
[7] base

other attached packages:
[1] FSA_0.8.4       extrafont_0.17 knitr_1.11
```

Acknowledgments

The following people shared their data for the book: Roger Harding (*CutthroatAL.csv*); Michael Hendricks and Richard McBride (*ShadCR.csv*); Joseph Hightower (*Zehfussetal_1985.INP*); Ryan Koenigs (*WaleyeWad.csv* and *WalleyeWyrlng.csv*); Ben Neely (*BGHRsample.csv* and *BGHRfish.csv*); Thomas Pratt (*SturgeonGB.csv*); Michael Quist (*CreekChub.csv*); the U.S. Fisheries and Wildlife Service Ashland Fish and Wildlife Conservation Office (*SiscowetMI2004.csv*); and Virginia Marine Resources Commission (*BlackDrum2001.csv*). Others contributed data that I ultimately did not use in the book. I am thankful to all of these people; their graciousness made the book much more practical and interesting.

I am also grateful to several people that provided clear explanations for how to conduct specific analyses in R. In particular, I thank Hans Gerritsen for describing the use of multinom() (Sections 5.2.3 and 5.5), Louis-Paul Rivest for assistance with results from **Rcapture** (Sections 9.3.2 and 9.4.2), Alexis Dinno for help with Dunn's test (Section 8.3.3), Torsten Hothorn for help understanding some of the inner workings of glht(), and Michelle C. Nelson for help with references for the Schnute method in removal() (Section 10.2).

Typesetting the book was made much easier with **knitr** and RStudio (RStudio 2015b). I thank Yihui Xie for his amazing work on **knitr**. This was, of course, not without some technical issues and I thank the good folks in the knitr Google Group and Stack Exchange communities, especially Yihui Xie, for their assistance with my questions. I also thank John Verzani for sharing the workflow he used to produce his *Using R for Introductory Statistics* book, Marcus Fontaine at Taylor & Francis for assistance with LaTeXquestions, and Linda Leggio at Taylor & Francis for editorial assistance.

Chapters of the book (at various stages of preparation) were reviewed by Michael Colvin (Chapters 9 and 10), Janice Kerns (Chapter 11), Young Lee (Chapter 2), Salvatore Mangiafico (Chapters 7, 8, and 12), Richard McBride (Chapter 4), Ben Neely (Chapters 1–3, 5–8, 11, and 12), Kaye Roncevic (Chapters 1–3), Brett Roper (Chapters 9 and 10), Michael Seider (Chapter 5), and Jason Smith (Chapter 11). Elise Zipkin and an anonymous reviewer provided larger picture comments on the entire book. Alan Brew and Abby Knoblauch offered valuable general editorial advice. I learned much from their thoughtful and constructive reviews that also led to substantive improvements in the book. Of course, any remaining errors or omissions are mine.

Books do not get written without a wide network of support for the author. I am fortunate to have had the support of David Grubbs at Taylor & Francis; the faculty, students, and administration of Northland College; many colleagues in the fisheries community; and my friends and family. I am thankful for your understanding for when I missed deadlines and declined invitations as I worked on this project.

My deepest thanks, however, go to Kim and Jem. Your unwavering faith in me and this project continuously inspired and motivated me. I cannot possibly thank both of you enough!!

Dedication

I have been fortunate enough to have had three significant mentors during my career. Nicholas Bystrom was my basketball coach and statistics and mathematics professor at Northland College. "Coach" helped me understand, very early in my career, the power, utility, and beauty of mathematics and statistics. James Selgeby was my boss for a work-study position with the U.S. Fish and Wildlife Service during my last two years at Northland College. Jim trusted me as a fisheries scientist, fostered my passion for Great Lakes fisheries, and encouraged me to pursue an advanced degree in fisheries. George Spangler was my doctoral advisor at the University of Minnesota. George strongly encouraged advanced studies in statistics, supported my interest in programming, and provided an environment, as a graduate student and later while on sabbatical, where I could creatively mix statistics, programming, and fisheries. This book would not exist without each of these great men. I dedicate this work to them.

Derek H. Ogle
Ashland, Wisconsin

Author

Derek Ogle is a Professor of Mathematical Sciences and Natural Resources at Northland College in Ashland, Wisconsin. Derek earned his Ph.D. in Fisheries Science from the University of Minnesota in 1996. He joined the faculty at Northland College in 1996 where he teaches statistics and fisheries science courses. His research interests depend on the interests of current students and collaborators but tend toward the population dynamics of invasive species and little studied native species. At Northland, Ogle has received awards for teaching, scholarly work, service, and assessment activities.

Ogle maintains the *fishR* website and blog (`http://derekogle.com/fishR`), Twitter account (@fishr_ogle), and Facebook page (`https://www.facebook.com/fishr00`), which are dedicated to sharing information on how to perform fisheries analyses in R.

1

(Very Brief) Introduction to R Basics

Specific foundational information about the R language and environment as it relates to the analysis of fisheries data is briefly introduced in this chapter. The intent here is to get you started with R so that you can proceed competently with later chapters. Much more thorough introductions to the R language and environment have been done well by many other authors (e.g., Dalgaard 2002; Fox and Weisberg 2011; Aho 2014; and Verzani 2014) and should be consulted for more details.

1.1 Why R for Fisheries Scientists?

The learning curve for the R language and environment is initially somewhat steep. However, the investment in learning R is worthwhile because R has several characteristics that will increase the creativity, rigor, and efficiency of your fisheries analyses. Four of these characteristics — use of the command line, near real-time extensions of functionality, open-source, and free cost — and their advantages are outlined below.[1]

The **command line interface** (examples shown below and throughout the book) requires the user to type instructions, rather than select options from a menu. This may seem like a disadvantage to those that have only used graphical user interfaces (GUI); however, I believe that there are two distinct advantages to using the command line. First, the command line forces the user to ask for particular results, which requires the scientist to be systematic, organized, and thoughtful about the analysis, both *a priori* and *a posteriori*. Second, instructions entered at the command line can be saved as a script (see Section 1.10). The script, along with the data and a description of the software setup (e.g., version numbers), can be shared with others who can then faithfully reproduce the analysis.[2] The ability to readily reproduce analyses likely leads to more precise communication of methods (exact analytical methods are known), increased scientific integrity (results can be verified both with the original data and new data), increased application of new analytical methods (steps for implementing the new method are provided), and increased collaboration (data and analytical steps can be easily shared among collaborators).

R is extensible in that new methods, either broadly within statistics or specific to an applied field (e.g., fisheries), can be delivered as they are developed. Indeed, R is continually being expanded by an ever-growing population of R users that create extensions (called *packages*; see Section 1.3) that add specific functionality to R. Often, these extensions are on the "cutting-edge" of a field.

R is open-source which broadly means that the source code for the software can be examined by the user.[3] The open-source nature of R leads to the extensions mentioned previously. To a particular user, however, open-source means that one can examine the exact instructions that underlie a particular analysis. Thus, analytical methods can be directly verified by the scientist performing the analysis and, ultimately, others.

Finally, **R is free**, which makes it available to a very large set of users. I have seen while administering the *fishR* website that R enables scientists in third world countries, cash-strapped agencies, or underfunded educational institutions to perform advanced analyses. The ability for all fisheries scientists to make management decisions from scientifically rigorous analyses is an overall benefit to global fish populations.

1.2 Installing R and RStudio

The R software is freely available from the Comprehensive R Archive Network (CRAN) at `https://www.r-project.org`. Instructions for downloading and installing R are available as a supplement to this chapter on the companion website.

RStudio is an integrated development environment (IDE) that "sits on top of" R. Among other things, it provides a text editor with syntax highlighting and seamless code execution with R. You can use other text editors with R, but RStudio is well organized and makes the interaction with R very efficient, especially for beginners. In addition, RStudio has the same appearance whether you use a Windows, Mac OS, or Unix/Linux platform. RStudio is available from `https://www.rstudio.com`. Instructions for downloading, installing, and configuring RStudio are given in the online supplement.

1.3 Packages

The functionality provided by the standard installation of R is referred to as either *base* or *core* R. Additional functionality is added to base R by installing and loading *packages*, which are a set of related functions (discussed more later) and data sets. For example, **plotrix** (Lemon 2006) provides functions to construct various plots that are not available in base R and **car** provides functions that support the book *An R Companion to Applied Regression* (Fox and Weisberg 2011). The **FSA** package supports fisheries analyses described in this book.

R packages are available for download from CRAN or from a number of other repositories (e.g., RForge or GitHub). Instructions for installing packages are given in the online chapter supplement.

A package needs to be installed only once (or whenever the package version changes), but must be loaded in each R session in which functions from the package will be used. Packages are loaded into the current R session by including the name of the package within `library()`. For example, the following code loads **FSA** into the current R session and makes its functions and data available for use.

```
> library(FSA)
```

If R is closed and then reopened (i.e., a new R session), then all needed packages must be reloaded with `library()` to make the functionality in the packages available for use during that R session.

Determining if a package exists to perform a particular analysis can be difficult. A set of "Task Views" is maintained at CRAN (`https://cran.r-project.org/web/views`) that can narrow the search to broad categories of topics. "CRANtastic" (`http://crantastic.org`) also provides the ability to search for packages available on CRAN. Packages related to fisheries analyses are collated on the fishR website (`http://derekogle.com/fishR/packages`). In this book, packages required to perform the analyses of a chapter are listed at the beginning of each chapter.

1.4 Prompts, Expressions, and Comments

In the R environment, very few, if any, instructions are accessed through menu options. Instead, instructional commands are entered at the *command prompt*, denoted by the > symbol.

```
>
```

The simplest R command is a *mathematical expression*, which is a simple instruction that tells R what to calculate. For example, the expression below illustrates the multiplication of two numbers.

```
> 2*6
[1] 12
```

The [1] in this result shows the position of the first item on each line of the result. In this case, there is only one item. The utility of this position label will be more apparent when working with vectors or data.frames (Section 1.7).

If an expression or command was not completed, then the *continuation prompt*, denoted with a +, will appear instead of the command prompt. The continuation prompt appears most often when quotes or parentheses were not closed. If the missing portion of the expression is at the end of the command, then the missing portion can be added at the continuation prompt to complete the code.

```
> (3+2)*(5+4
+ )
[1] 45
```

However, if the missing portion is not at the end of the command, then the correction cannot be made at the continuation prompt. In these instances, press the ESCape key while the cursor is in the console window to return to the command prompt and then correct and rerun the offending expression or command. To keep the code more readable, continuation prompts will not be shown in this book.

Any "code" following a "pound sign" or "hashtag" symbol (#) will not be interpreted by R. This provides a simple method to add comments to your R commands.

```
> # This is a whole-line comment
> 3+2*7  # This is a post-line comment
[1] 17
```

Extensive use of comments can provide directions to others or a reminder to you of what a command does. The use of comments within a script is illustrated in Section 1.10.

1.5   Objects

Everything in R is an object. However, for the purposes of this book an *object* is a name that refers to something (e.g., a value or data.frame) stored in memory for later use. A result is assigned to an object with the *assignment operator* (<-).[4] The expression or command is to the right of the assignment operator and a name for the object is to the left. Object names may NOT begin with numbers or contain spaces, colons, semicolons, mathematical operators (i.e., +, -, *, /, ^, =), or "special characters" (i.e., #, ?, %, $, &, |, !). Object names are case-sensitive; thus, result, Result, and RESULT are different objects.

The result from the simple expression from above is assigned to an object named `tmp` below.

```
> tmp <- 2*6
```

This example illustrates that the result is not displayed when it is assigned to an object. The result is seen, however, by subsequently typing the name of the object.

```
> tmp
[1] 12
```

Additionally, an extra set of parentheses around the command at the time of assignment will cause the result to both be assigned to the object and displayed in the console window.

```
> ( tmp <- 7+4*2 )
[1] 15
```

As illustrated in this last example, the result in an object is replaced when another result is assigned to the same name.

Objects can be used in expressions (or functions, see next section).

```
> tmp*3
[1] 45
```

1.6   Functions

Functions are programs (sets of commands) that perform a specific task. Functions are called with the function name followed immediately by parentheses. Within the parentheses are *arguments* that either provide information or directives to the function. For example, the function named `log10()` computes the common logarithm (i.e., base 10) for the value provided in its only argument.

```
> log10(1000)
[1] 3
```

Some functions allow more than one argument, which are separated by commas within the parentheses. Arguments may be declared by their position within the list of arguments or by an argument name. Some arguments are set at default values and, thus, are optional. Required arguments are usually the first few arguments to the function, whereas optional arguments are usually (but not necessarily) declared by name.

For example, `log()` will compute the logarithm to any base for the value supplied as its first argument. A second argument, named `base=`, defaults to `exp(1)` (i.e., R's representation of e, the natural constant). Thus, if `base=` is not explicitly declared, either by name or with a value in the second argument position, then `log()` will return the natural logarithm. However, if `base=` is set equal to some value, or a value is in the second argument position, then the logarithm to that base is returned.

```
> log(1000)                 # base= set by default (to exp(1))
[1] 6.907755
> log(1000,base=10)         # base= set explicitly, by name
[1] 3
> log(1000,10)              # base= set explicitly, by position
[1] 3
```

If more than two named arguments are being used, then they can appear in any order after the unnamed arguments. For example, `format()` is used to control the display of numbers. In the example below, `format()` is directed to display the number given in the first argument in scientific format (i.e., `scientific=TRUE`) with four significant digits (i.e., `digits=4`). Note how the order of the two named arguments is irrelevant.

```
> num <- 123456789
> format(num,scientific=TRUE,digits=4)
[1] "1.235e+08"
> format(num,digits=4,scientific=TRUE)
[1] "1.235e+08"
```

Argument names, their meaning, and their default values are seen in the help documentation for the function (see Section 1.11). Functions are discussed further in Section 1.8

1.7 Data Storage

1.7.1 Vectors or Variables

The simplest data structure in R is the *vector*, which is a one-dimensional array of elements that are all of the same data type. Data types in R can be numeric,[5] character (or string), logical (i.e., TRUE or FALSE), or factors (see Section 1.7.2). Vectors can be created by combining together elements of the same type with `c()`. For example, four vectors with six items each are constructed below.

```
> ( lens <- c(75,87,45,63,77,93) )
[1] 75 87 45 63 77 93
> ( wts <- c(13,14.2,7.8,9,10.1,10.1) )
[1] 13.0 14.2  7.8  9.0 10.1 10.1
> ( sex <- c("M","F","F","M","M","F"))
[1] "M" "F" "F" "M" "M" "F"
> ( mat <- c(TRUE,TRUE,FALSE,FALSE,FALSE,TRUE) )
[1]  TRUE  TRUE FALSE FALSE FALSE  TRUE
```

The type of data in a vector is seen with `class()`.

```
> class(lens)
[1] "numeric"
> class(wts)
[1] "numeric"
> class(sex)
[1] "character"
> class(mat)
[1] "logical"
```

Elements in a vector may also be named.

```
> ( hab <- c(rocky=35,sandy=45,silty=20))
rocky sandy silty
   35    45    20
```

Specific elements within a vector are accessed by including the numerical or named position of the element within [] following the object name. For example, the following code extracts the third element in the `lens` object.

```
> lens[3]
[1] 45
```

Multiple elements are returned by combining positions into a vector with `c()` and then using that vector within [].

```
> lens[c(1,3)]
[1] 75 45
```

Elements are excluded by preceding the position(s) with a negative sign.

```
> lens[-c(1,3)]
[1] 87 63 77 93
```

For named vectors, the elements may be accessed by position or by name (the name must be in quotes).

```
> hab[2]
sandy
   45
> hab["sandy"]
sandy
   45
```

At times, it may be necessary to remove the names from the resulting vector. For example, the name may be inappropriate if a mathematical operation is performed on the result. In these instances, use `[[]]` rather than `[]` to extract the element.

```
> 100-hab[2]         # old name is inappropriate
sandy
   55
> 100-hab[[2]]       # no name
[1] 55
```

However, `[[]]` **ONLY** works if a single element is returned.

```
> hab[[c(2,3)]]
Error in hab[[c(2, 3)]]: attempt to select more than one element
```

As seen above, mathematical operations can be performed with numeric vectors. For example, each value in a vector may be multiplied by a constant, have a constant value added to it, or raised to a constant exponent.

```
> lens/25.4
[1] 2.952756 3.425197 1.771654 2.480315 3.031496 3.661417
> lens+10
[1]  85  97  55  73  87 103
> lens^3
[1] 421875 658503  91125 250047 456533 804357
```

Alternatively, operations are performed element-by-element for vectors with the same number of elements.

```
> lens+wts
[1]  88.0 101.2  52.8  72.0  87.1 103.1
> wts/(lens^3)*10000
[1] 0.3081481 0.2156406 0.8559671 0.3599323 0.2212326 0.1255661
```

Mathematical operations on vectors are subject to *recycling*, where a shorter vector is repeated as many times as necessary to match the size of the longer vector. For example, if one attempts to sum a vector with six elements and a vector with two elements, then the vector with two elements is repeated three times.

```
> lens + c(10,100)
[1]   85 187  55 163  87 193
```

A warning is given if the shorter vector is not a multiple of the longer vector. The operation, however, will be completed as shown below.

```
> lens + c(1,10,100,1000)
Warning in lens + c(1, 10, 100, 1000): longer object length is
not a multiple of shorter object length
[1]   76   97 145 1063   78 103
```

Recycling can be advantageous (this is what allows mathematical operations with a "constant value", which is simply a vector with one element), but it can produce unintended results if the user is not careful. Thus, one should get in the habit of checking the result after every manipulation.

1.7.2   Factors

The fisheries scientist may have character data that should be treated as categorical data where each "word" represents a category or group to which an individual belongs. In R, categorical data are stored as a *factor*. A character vector is coerced to be a factor vector with `factor()`.

```
> ( sex1 <- factor(sex) )
[1] M F F M M F
Levels: F M
> class(sex1)
[1] "factor"
```

By default, `factor()` will order the levels alphabetically. There may be times when the alphabetical order of the levels does not make sense, such as when the month of capture (i.e., "Jan", "Feb", etc.) or sediment grain size category (i.e., "Boulder", "Cobble", "Pebble", etc.) was recorded. In this or other situations, such as when statistical analyses (e.g., analysis of covariance or logistic regression) treat the first level as a reference group, the alphabetical order of levels may be undesirable. The order of levels for a factor is changed by entering the level names, in the desired order, into a character vector and then submitting that vector to `levels=` in `factor()`.[6]

```
> ( sex2 <- factor(sex,levels=c("M","F")) )
[1] M F F M M F
Levels: M F
```

It is important to note, as illustrated in the examples with `sex1` and `sex2`, that changing the order of the levels does not change the actual data, it simply changes which level is considered "first."

"Behind-the-scenes" R uses numeric codes for the levels, starting at 1 for the first level and continuing sequentially to the last level. While not usually necessary, these codes can be seen by submitting the factor variable to as.numeric().

```
> as.numeric(sex1)
[1] 2 1 1 2 2 1
> as.numeric(sex2)
[1] 1 2 2 1 1 2
```

Unless explicitly coerced to numeric values, as with as.numeric(), the user will see the "words" rather than the numeric codes associated with each level of a factor variable. However, the numeric codes underlying the factor variables can be exploited, as shown in Sections 3.1.2 and 12.4.

1.7.3 Matrices

Matrices are two-dimensional arrays (i.e., both rows and columns) of the **same** data type. In this book, matrices will only be constructed by binding vectors of the same data type.[7] Specifically, vectors are *row-bound* with rbind() or *column-bound* with cbind().

```
> rbind(lens,wts)
     [,1] [,2] [,3] [,4] [,5] [,6]
lens   75 87.0 45.0   63 77.0 93.0
wts    13 14.2  7.8    9 10.1 10.1
> ( lw <- cbind(lens,wts) )
     lens  wts
[1,]   75 13.0
[2,]   87 14.2
[3,]   45  7.8
[4,]   63  9.0
[5,]   77 10.1
[6,]   93 10.1
```

Matrices **must** have all elements of the same type. If one tries to bind vectors of different data types, then the result is reduced to be only one of the data types.[8]

```
> rbind(lens,sex)
     [,1] [,2] [,3] [,4] [,5] [,6]
lens "75" "87" "45" "63" "77" "93"
sex  "M"  "F"  "F"  "M"  "M"  "F"
```

```
> rbind(lens,mat)
     [,1] [,2] [,3] [,4] [,5] [,6]
lens   75   87   45   63   77   93
mat     1    1    0    0    0    1
> rbind(sex,mat)
     [,1]    [,2]    [,3]     [,4]     [,5]     [,6]
sex  "M"     "F"     "F"      "M"      "M"      "F"
mat  "TRUE"  "TRUE"  "FALSE"  "FALSE"  "FALSE"  "TRUE"
```

Thus, you should examine any data structure after you have constructed it to make sure that the result is what you expected.

Elements in a matrix are also accessed with [] or [[]]. However, for a matrix, the positions must be a row- and column-number pair separated by a comma.

```
> lw[[1,2]]            # 1st row, 2nd column (no name for element)
[1] 13
> lw[c(1,3),2]         # 1st & 3rd rows, 2nd column
[1] 13.0  7.8
> lw[-c(1,3),2]        # exclude 1st & 3rd rows, 2nd column
[1] 14.2  9.0 10.1 10.1
```

All rows or columns are returned if that position is left blank within [].[9] The comma inside the square brackets must be included because it acknowledges that the two-dimensional matrix requires a row and column position.

```
> lw[1,]                              # 1st row
lens   wts
  75    13
> lw[,1]                              # 1st column
[1] 75 87 45 63 77 93
> lw[,"wts"]                          # wts column
[1] 13.0 14.2  7.8  9.0 10.1 10.1
```

1.7.4   data.frames

Vectors are limiting because they are one-dimensional and matrices are limiting because all entries must be of the same data type. Data.frames, however, are two-dimensional and can have **columns** of different data types. Thus, data.frames are useful for storing real data that often consists of multiple types of variables.

Existing vectors are combined into a data.frame with **data.frame()**.

```
> ( df <- data.frame(lens,wts,sex,mat) )
  lens  wts sex    mat
1   75 13.0   M   TRUE
2   87 14.2   F   TRUE
3   45  7.8   F  FALSE
4   63  9.0   M  FALSE
5   77 10.1   M  FALSE
6   93 10.1   F   TRUE
```

It is not efficient, however, to enter variables recorded for many individuals into separate vectors in R and then create the data.frame with `data.frame()`. Thus, real data are usually entered into an external database (or spreadsheet) and then loaded into a data.frame in R. Loading data from an external file into a data.frame is demonstrated in Section 2.1.

Columns in data.frames correspond to variables, whereas rows correspond to individuals.[10] Elements of a data.frame are also extracted with [].

```
> df[1,2]                                # 1st row, 2nd column
[1] 13
> df[1,]                                 # 1st row/individual
  lens wts sex  mat
1   75  13   M TRUE
> df[,2]                                 # 2nd column/variable
[1] 13.0 14.2  7.8  9.0 10.1 10.1
> df[,"wts"]                             # wts column/variable
[1] 13.0 14.2  7.8  9.0 10.1 10.1
```

However, single variables (i.e., columns) are extracted from a data.frame more efficiently with the $ operator. Specifically, `df$var` extracts the generic variable `var` from the generic data.frame `df`.

```
> df$lens
[1] 75 87 45 63 77 93
> df$sex
[1] M F F M M F
Levels: F M
```

Each variable extracted from a data.frame is a vector. Thus, all operations that can be performed on a vector can be performed on a variable from a data.frame.

```
> class(df$lens)
[1] "numeric"
> df$lens[c(1,3)]
[1] 75 45
```

```
> df$lens/25.4
[1] 2.952756 3.425197 1.771654 2.480315 3.031496 3.661417
```

Thus, it is common to extract a variable from a data.frame to be used as a vector argument to a function. For example, the code below computes the common logarithm of all lengths in the `lens` variable of the `df` data.frame.

```
> log10(df$lens)
[1] 1.875061 1.939519 1.653213 1.799341 1.886491 1.968483
```

Other functions, however, allow the variable to be identified with *formula notation* and `data=`. Formula notation uses the ~ operator and is illustrated briefly in the next section and is used throughout this book.

1.8   More with Functions

1.8.1   Constructor and Extractor Functions

A general philosophy of R is to make the user ask for specific results. With this philosophy, some R functions will perform many calculations and store the results into a data structure (i.e., vector, data.frame, matrix, or list). The user must then use other functions to extract specific pieces of information from that data structure.

For example, it is common for fisheries scientists to perform a simple linear regression to find the best-fit line for two quantitative variables. The process of finding the best-fit line (described in many introductory statistics textbooks) is accomplished with `lm()`. The use of `lm()` is described in detail in Section 7.3; however, here, we simply need to know that the first argument is a formula of the form y~x and the data.frame that contains the quantitative y and x variables is given in `data=`. For example, the code below "performs" a simple linear regression between the `wts` and `lens` variables in the `df` data.frame created in the previous section.

```
> lm1 <- lm(wts~lens,data=df)
```

The `lm()` function is called a *constructor function* because it creates a great deal of information that it stores in a data structure. The results of constructor functions are almost always assigned to an object, as done above with `lm1`, from which further information is extracted or analyses are performed. The results in this object are seen, but are not shown here, by submitting the object to `str()`.

Specific information is extracted from the object created by the constructor function with an *extractor function*. For example, the y-intercept and slope for the best-fit model are extracted with `coef()`, the residuals from the model fit are extracted with `residuals()`, and an ANOVA table is extracted with `anova()` (these results are discussed in detail in Section 7.3).

```
> coef(lm1)
(Intercept)          lens
 3.97967986   0.09164073

> residuals(lm1)
          1          2          3          4          5          6
   2.147265   2.247577  -0.303513  -0.753046  -0.936016  -2.402268

> anova(lm1)
Analysis of Variance Table

Response: wts
          Df Sum Sq Mean Sq F value Pr(>F)
lens       1 12.591 12.5914  2.9682   0.16
Residuals  4 16.969  4.2421
```

One need not be too worried about the terms "constructor" and "extractor." However, you will see in subsequent chapters that it is common to submit the results of one function (a constructor function) to another function (an extractor function) to return specific results. Fortunately, extractor functions are quite general. For example, `coef()` returns the parameter estimates from many types of models, not just those returned by `lm()`.

1.8.2   User-Defined Functions

R contains a programming language that allows users to define their own functions. Defining your own function may minimize errors, reduce repetitive calculations, and allow implementation of new or complex procedures. For example, new functions are created in Sections 12.3.2 and 13.1.3 to make complex calculations more efficient. A general procedure to create a new function is illustrated with a simple example below.

Suppose you want to use the best-fit line from the previous section to predict the weight of an individual fish given its length. This is accomplished by multiplying the length of the fish by the slope and adding the intercept. To illustrate this calculation, both the intercept and slope are extracted from the `lm()` object below and assigned to objects.

```
> ( int <- coef(lm1)[[1]] )
[1] 3.97968
> ( slp <- coef(lm1)[[2]])
[1] 0.09164073
```

The length of the fish is then assigned to an object before it is entered into an expression to predict the weight.

```
> len <- 75                    # a chosen length
> ( pred.wt <- slp*len+int )   # the predicted weight
[1] 10.85273
```

This is cumbersome code[11] that is easily subject to error and inefficiency if it is to be repeated. In instances like this, it may be useful to wrap this code into a function that will produce the desired end result more reliably and efficiently.

New functions are created with `function()` followed by a set of "curly brackets" that contains the body (i.e., code) of the function. The body of the function contains the code that performs a task. The result is returned in the last line of the body of the function. The arguments to `function()` become the arguments of the newly created function. The result of `function()` is assigned to an object which is then the name of the new function.

For example, the code below creates a new function, `predY`, that takes the value at which to make the prediction as its first argument (`x`), the best-fit linear model object as its second argument (`lmobj`), and returns the predicted value. The body of the function is nearly the same as the code used above, except that within the function, the given length is in `x` and the linear model results are in `lmobj`.

```
> predY <- function(x,lmobj) {
    # get y-intercept and slope
    int <- coef(lmobj)[[1]]
    slp <- coef(lmobj)[[2]]
    # make the prediction
    pred.wt <- slp*x+int
    # return the prediction
    pred.wt
  }
```

The first line below calculates the same prediction demonstrated previously. Furthermore, the second line demonstrates how the function is used to predict weights for the vector of lengths in `df$lens`.

```
> predY(75,lm1)        # single prediction (compare to previous)
[1] 10.85273
> predY(df$lens,lm1)   # vector of predictions
[1] 10.852735 11.952423  8.103513  9.753046 11.036016 12.502268
```

In this book, the need for user-defined functions is limited to only a few very simple functions that return only a single value or a vector of values (see Sections 12.3.2 and 13.1.3.3). As illustrated here, your own functions can make very repetitive code easier to read and more efficient to calculate. Thus, writing your own functions is a skill that is worth developing as you become more familiar with R. For the interested reader, function creation is described in more detail in Fox and Weisberg (2011) and Wickham (2014a).

1.9   Looping

1.9.1   For Loops

Some calculations require the user to iteratively insert a variety of values into an expression or function. These iterations may be accomplished with a loop that steps through the values. Loops are created with `for()`, which takes an argument of the form `var in seq`, where `var` is a variable used in the loop and `seq` is a sequence of values that that variable will take during each step of the loop. For example, `for(i in 1:5)` will result in i <- 1, then i <- 2, and so on through i <- 5. The code to be executed at each step of the loop follows `for` and is contained within "curly brackets" if the code is more than one line.

The result of the loop is usually assigned to an object. For the sake of efficiency, an object of the appropriate type and size should be created before the loop begins. For this purpose, an "empty" numeric vector is constructed with `numeric()` and a character vector with `character()`. In these functions, the only argument is the number of items to be in each vector (i.e., the length of each vector). "Empty" matrices are created with `matrix()`, with the number of rows given in `nrow=` and the number of columns in `ncol=`.

As a simple example, the loop below creates a vector that contains the squares of the values in `tmp`.[12] Note that `length()` returns the number of items in a vector. In this example, the result is printed at each step of the loop so that you can see how the result vector is sequentially filled. One would not usually print the result after each step.

```
> # create vector of values
> tmp <- c(7,2,3,5,4)
> # use loop to find the square of each value
> res <- numeric(length(tmp))
> for(i in 1:length(res)) {
    res[i] <- tmp[i]^2
    print(res)
  }
[1] 49  0  0  0  0
[1] 49  4  0  0  0
[1] 49  4  9  0  0
[1] 49  4  9 25  0
[1] 49  4  9 25 16
```

As a more substantive example, the following loop computes the total number of fish caught in each of three successive sampling events (columns) at two locations (rows) stored in a matrix. The code initializes the vector to hold the results (note that `nrow()` returns the number of rows in its argument), iteratively sends each row of the matrix to `sum()`, and stores the result in the appropriate position of the vector.[13]

```
> # create a matrix
> m <- rbind(c(15,6,2),c(30,15,5))
> # find the sum of the rows of that matrix
> res <- numeric(nrow(m))
> for(i in 1:length(res)) res[i] <- sum(m[i,])
> res
[1] 23 50
```

More realistic examples are given in Sections 12.3.5 and 13.2.2. A more thorough introduction to loops is given in Matloff (2011) and Aho (2015b).

1.9.2 Apply "Loops"

The apply family of functions make common looping problems more efficient, both in terms of minimizing code and speed of processing. These functions include `apply()`, `lapply()`, `tapply()`, and several others. Only `apply()` is demonstrated below.

The `apply()` function is used to submit the rows or columns of a matrix to the first argument of another function. For example, the `apply()` below submits each row of the matrix created in the previous section to `sum()` (which computes the sum of the values in the vector given to it). The first argument to `apply()` is the matrix. This matrix is processed by rows if `MARGIN=1` and by columns if `MARGIN=2`. The function to be applied to each row or column is given in `FUN=`. Other arguments required by the function in `FUN=` are given by name as arguments to `apply()` (which is not illustrated here).

```
> ( res1 <- apply(m,MARGIN=1,FUN=sum) )   # by rows
[1] 23 50
> ( res2 <- apply(m,MARGIN=2,FUN=sum) )   # by columns
[1] 45 21  7
```

More realistic uses of `apply()` are in Sections 6.3.5, 10.2, and 12.3.4. The learning curve for the apply family of functions can be steep. More detailed explanations and examples are in Fox and Weisberg (2011), Matloff (2011), and Aho (2015b).

1.10   Saving Results

The results in objects are not saved when exiting R, unless you choose to save
the workspace. For beginners, it is best not to save the workspace. Rather, it
is better to save your commands as a script so that you can run the script
again to reproduce the analysis.

A script is a simple text file (usually with a ".R" extension) of R commands,
usually annotated with comments. The code excerpt below is an example of a
simple script. Scripts that contain the code used in each chapter of this book
are available on the companion website for the book.

```
> ## This an example script that creates very simple data and
> ##    computes the mean length and weight.
>
> ## Create simple data
> df <- data.frame(lens= c(75,87,45,63),
                   wts= c(13,14.2,7.8,9),
                   sex <- factor(c("M","F","F","M")),
                   mat <- c(TRUE,TRUE,FALSE,FALSE) )
> ## compute the mean length and weight
> mean(df$lens)
> mean(df$wts)
```

1.11   Getting Help

R provides several resources for getting help about certain functions. If you
know the name of a function, the help documentation is retrieved by preceding
the function name with the ? operator. For example, ?mean will display the
help documentation for mean().

There will be times, however, when part of the help you seek IS the name
of a function. In these cases, you can search for keywords in the help documentation for all functions in all installed packages on your system by placing
the keyword, within quotes, as the argument to help.search(). For example,
one can find all functions in installed packages that use the word "mortality"
with help.search("mortality").

One can do a similar search, as long as you have an Internet connec-
tion, in the help documentation of all functions in base R and all packages
on CRAN with RSiteSearch(). Again, the first argument is a keyword. Including restrict="functions" will limit the results to those results related
to R functions (rather than, for example, also from data files). The results

are returned in your web browser. For example, one can find all functions in base R and CRAN packages that use the word "Mark-Recapture" with `RSiteSearch("Mark-Recapture",restrict="functions")`.

Quality answers to many questions may also be found by searching the Internet, especially the archives of the "R-Help" mailing list (`http://tolstoy.newcastle.edu.au/R/`) and StackOverflow (`http://stackoverflow.com/questions/tagged/r`). Furthermore, an R FAQ is maintained at `https://cran.r-project.org/faqs.html`.

Finally, **as a last resort**, one can pose good questions to the "R-help" mailing list or, with an R tag, to StackOverflow. Please pay close attention to the posting guide for R-Help (`https://www.r-project.org/posting-guide.html`). Additionally, for both R-Help and StackOverflow, your questions should include a minimal, self-contained, reproducible example. A guide to producing such an example can be found by searching for "great R reproducible example" on StackOverflow.

Notes

[1] One can perform an Internet search on "Why use R" to find characteristics of R that others deem important.

[2] What I called "reproducible analysis" is often called "reproducible research" by others. The concept of reproducible research has gained great favor in the last decade. While reproducible research means different things to different people, the core of the idea is that data analyses can be readily reproduced. The reproducible research literature has become vast, but useful entry points are Peng et al. (2006) and Gandrud (2014).

[3] The source code for many functions is obtained by typing the name of the function without any following parentheses or arguments. Some functions appear hidden to the user, but these functions can often be seen by including the name of the function within quotes in `getAnywhere()`. Some other functions are more difficult to see the source code because what the user types as the function name is not the actual function name used. For example, the user might type `summary()`, but the actual function used depends on the arguments to `summary()`. For example, if the argument to `summary()` is a data.frame, then `summary.data.frame()` is actually called. The source code for `summary.data.frame()` is available by typing the function name without parentheses. The actual function names (called S3 methods) that may underlie a general function name are seen by including the name of the function in `methods()`.

[4] Some users will use an equals sign for the assignment operator. There are pages of arguments for and against both operators on various Internet sites. My conclusions from reading these arguments is that there are some instances where using = instead of <- as the assignment operator may cause problems, but these instances are likely to be very rare for the practicing fisheries scientist and at the introductory level of this book. Despite being somewhat longer to type, I use <- throughout this book because I do not want to risk an occurrence of the rare problems caused by =, I like the look of <- (especially with preceding and succeeding spaces), and code like `t <- t+1` feels legitimate to me, whereas `t = t+1` does not.

[5] Numeric data may be stored as either integers or doubles. One can force the data to be an integer by appending an L to the number. For example, `c(3L,4L)` will produce a vector with two integer values.

[6]The `relevel()` function is used to move one level to the first position with the remaining levels maintaining their same relative order. The first argument to `relevel()` is the factor vector to be reordered and the second argument is the name of the level, in quotes, that should be moved to the first position of the levels.

[7]Matrices may also be constructed with `matrix()`.

[8]When different types of vectors are given to `rbind()` or `cbind()`, R uses a hierarchical scheme to determine the data type for the resulting matrix. That scheme is described in the help documentation for either binding function. In this book, we will only bind vectors of the same data type.

[9]The `[[]]` cannot be used in this context because more than one element (e.g., an entire row, or vector) is returned.

[10]Individual is used here to denote an item that had data recorded about it. Synonyms used by other authors are unit, experimental unit, observational unit, or subject.

[11]This code could have been written in one line. In addition, the predictions may be made with `predict()` as described in Section 7.3.1. This code was used here to be very deliberate, but also to illustrate how a function can simplify seemingly complex calculations.

[12]This example is more efficiently computed with `tmp^2`.

[13]This example is more efficiently computed with `rowSums()` (e.g., `rowSums(m)`).

2

Loading Data and Basic Manipulations

Data are at the center of most fisheries analyses. Various data manipulations, such as adding or deleting variables, filtering or sorting rows according to particular criteria, converting to a different format, combining vertically or via a key value variable, or forming new data by aggregating across groups may be required before performing certain analyses. This chapter describes how to load data into R and perform the various manipulations commonly required by fisheries scientists to perform the analyses in subsequent chapters.

Required Packages for This Chapter

Functions used in this chapter require loading the packages shown below.

```
> library(FSA)
> library(magrittr)
> library(dplyr)
> library(tidyr)
```

2.1   Loading Data into R

Fisheries data are most often *rectangular*, where variables form the columns, observations on individuals or cases form the rows, and each row contains an entry for each column (Wickham 2014b). For example, if seven variables were recorded on 676 fish, then the corresponding file would have seven columns and 676 rows. Data files are often augmented with the names of the variables in a *header* (i.e., the first row above the actual data). Figure 2.1 shows the header and first ten rows of a rectangular data file.

Fisheries data are entered directly into R only when a few variables on a small number of individuals exist (see Section 1.7.4). Fisheries data are most often entered into a flat (e.g., spreadsheet) or relational database, but may also be available in the format of other statistical software (e.g., SAS). Methods for loading data from these external files into a data.frame in R is described in this section.

netID	fishID	species	tl	w	tag	scale
12	16	Bluegill	61	2.9		FALSE
12	23	Bluegill	66	4.5		FALSE
12	30	Bluegill	70	5.2		FALSE
12	44	Bluegill	38	0.5		FALSE
12	50	Bluegill	42	1		FALSE
12	65	Bluegill	54	2.1		FALSE
12	66	Bluegill	27			FALSE
13	68	Bluegill	36	0.5		FALSE
13	69	Bluegill	59	2		FALSE
13	70	Bluegill	39	0.5		FALSE

FIGURE 2.1. A header row (i.e., first row with variable names) and the first ten rows of data for the rectangular *inchBio.csv* data file as it would appear in a spreadsheet.

2.1.1   Working Directory

Functions used to load data from external files search for those files in the *working directory* of R, which is seen with getwd().

```
> getwd()
[1] "C:/aaaWork/Research/IFARbook/2mmatter/DataManip"
```

If the current working directory does not contain the file, then the directory must be changed (or "set") to where the file is located with setwd().[1] The only argument to setwd() is the full path to the directory, in quotes and using either the forward slash (/) or double backslashes (\\) to separate levels of folders.[2]

```
> setwd("C:/data")
```

2.1.2   Database and Other Software Files

Fisheries data are often entered into a database (including spreadsheets), which can be loaded directly into R. However, to do so may require configuration of software other than R (e.g., Java, PERL) that are specific to your operating system.[3] Data recorded in a database are often exported to a text file[4] which is then loaded into R as described in Section 2.1.3.

Readers interested in loading data directly from a database should examine **gdata** (Warnes et al. 2015), **readxl** (Wickham 2015), **RODBC** (Ripley and Lapsley 2015), **XLConnect** (Mirai Solutions GmbH 2015), and **xlsx** (Dragulescu 2014) for Microsoft Excel files; **RODBC** for Microsoft Access files; and **RMySql** (Ooms et al. 2015), **RODBC**, **ROracle** (Mukhin et al. 2014), **RPostgreSQL** (Conway et al. 2013), and **RSQLite** (Wickham et al. 2014) for various forms of SQL database files. Additionally, see **foreign** (R Development Core Team 2015a), **haven** (Wickham and Miller 2015), or **Hmisc** (Harrell 2015) for functions to load SAS or SPSS files and **foreign** to load Minitab, Octave, and Systat files. Finally, the relatively new **rio** (Chan et al. 2015) appears capable of reading data from a variety of formats with the same function.

2.1.3   Text or ASCII Files

Text or ASCII files are very simple files that are readable by most text editors (e.g., Notepad), word processors, and spreadsheet software. In a text file, the columns of data are separated by a *delimiter*, which is most often one or more spaces, a tab, or a comma. A *tab-delimited text file* uses tabs and a *comma-separated values file* (CSV) uses commas to separate observations within a row.

Text files are smaller and more portable than database files and are likely to be read by all future versions of R and other software. A new text file, however, will need to be created if the original data were changed in the database. In other words, the user runs the risk when using text files of having two versions of the data — one in the original database file and one in the text file. If the original database is a spreadsheet, then this risk is mitigated by using the CSV format, which is interpreted seamlessly by nearly all spreadsheet software. Thus, for most fisheries applications, the advantages likely out-weigh the disadvantages of using a text file, especially one in CSV format. Thus, CSV files are used throughout this book.

The `inchBio.csv` file contains the unique net identification number, unique fish identification number, species, total length (mm), weight (g), tag "number," and whether scales were sampled or not for individual fish captured in 2007 and 2008 from Inch Lake (a small catch-and-release only lake in northern Wisconsin). These data were originally entered into a relational database, but were exported to a CSV file with variable names in the first row (Figure 2.1). The header and first 10 rows of data in the CSV file look like the following in a text editor.

```
netID,fishID,species,tl,w,tag,scale
12,16,Bluegill,61,2.9,,FALSE
12,23,Bluegill,66,4.5,,FALSE
12,30,Bluegill,70,5.2,,FALSE
12,44,Bluegill,38,0.5,,FALSE
12,50,Bluegill,42,1,,FALSE
12,65,Bluegill,54,2.1,,FALSE
12,66,Bluegill,27,,,FALSE
13,68,Bluegill,36,0.5,,FALSE
13,69,Bluegill,59,2,,FALSE
13,70,Bluegill,39,0.5,,FALSE
```

Both CSV and tab-delimited text files are loaded into a data.frame in R with `read.table()`, with the name of the external file in quotes as the first argument. The type of separator for the columns is given in `sep=`, with `sep="\t"` for tabs and `sep=","` for commas. If the first row of the file is a header, then `header=TRUE` is included. By default, `read.table()` treats `NA` values in the data file as *missing*. This behavior is changed by including the string to be considered as missing in `na.strings=`. For example, use `na.strings=""` to code "blanks" as missing. Other arguments to `read.table()` sometimes used by fisheries scientists are `dec=`, `skip=`, and `stringsAsFactors=`. The help documentation for `read.table()` should be consulted for use of these arguments. The results of `read.table()` are assigned to an object, which becomes the name of the data.frame in R.

```
> bio <- read.table("inchBio.csv",sep=",",header=TRUE)
```

Comma-separated values files are loaded more succinctly with `read.csv()`, which is a wrapper function for `read.table()` that defaults to using `sep=","` and `header=TRUE`.

```
> bio <- read.csv("InchBio.csv")
```

The data.frame object should be examined immediately after it has been created. The *structure* of the data.frame is seen with `str()`, as shown below. The structure shows the number of observations or rows (676) and variables or columns (7) in the data.frame and the names (following $), data type (or class), and the first few values of each variable. For example, `netID` is the first variable, it is an integer type, and the first three values are 12, 12, and 12.

```
> str(bio)
'data.frame': 676 obs. of  7 variables:
 $ netID  : int  12 12 12 12 12 12 12 13 13 13 ...
 $ fishID : int  16 23 30 44 50 65 66 68 69 70 ...
 $ species: Factor w/ 8 levels "Black Crappie",..: 2 2 2 2 2 2..
 $ tl     : int  61 66 70 38 42 54 27 36 59 39 ...
 $ w      : num  2.9 4.5 5.2 0.5 1 2.1 NA 0.5 2 0.5 ...
 $ tag    : Factor w/ 193 levels "","1014","1015",..: 1 1 1 1 ..
 $ scale  : logi  FALSE FALSE FALSE FALSE FALSE FALSE ...
```

The structure of `bio` shows that `species` is a factor variable that contains eight levels, one of which is Black Crappie. The other levels for a factor variable are seen by including that variable within `levels()`.[5]

```
> levels(bio$species)
[1] "Black Crappie"    "Bluegill"         "Bluntnose Minnow"
[4] "Iowa Darter"      "Largemouth Bass"  "Pumpkinseed"
[7] "Tadpole Madtom"   "Yellow Perch"
```

One could view the entire data.frame by typing the name of the data.frame object.[6] However, it is usually more prudent to examine only a few rows of the data.frame. For example, the first and last six rows of the data.frame are shown with head() and tail(), respectively. A total of six rows from both the head and the tail (evenly distributed) is seen with headtail() from **FSA**. More or fewer rows may be seen with n= in each of these functions.

```
> headtail(bio,n=4)
    netID fishID       species  tl     w  tag scale
1      12     16      Bluegill  61   2.9      FALSE
2      12     23      Bluegill  66   4.5      FALSE
3      12     30      Bluegill  70   5.2      FALSE
4      12     44      Bluegill  38   0.5      FALSE
673   121    812 Black Crappie 142  37.0       TRUE
674   110    863 Black Crappie 307 415.0 1783  TRUE
675   129    870 Black Crappie 279 344.0 1789  TRUE
676   129    879 Black Crappie 302 397.0 1792  TRUE
```

2.1.4   Internal Files

Base R and many packages contain *internal* data which are loaded with data(). For example, the *Mirex* data.frame from **FSA** is loaded below.

```
> data(Mirex)
> tail(Mirex,n=3)
    year weight mirex species
120 1999   9.54  0.11 chinook
121 1999  11.36  0.09 chinook
122 1999  11.82  0.09 chinook
```

Internal data are primarily used to demonstrate functions in base R or a package and are not used with your own data or data provided in an external data file. Thus, you will likely only encounter data() in help documentation examples (Section 1.11) or in publications describing the features of a particular package.

2.2   Basic Data Manipulations

Data manipulations that can be applied to a single data.frame are illustrated with the bio data.frame in this section. The results from all data manipulations described below may be assigned to a new data.frame object with <-, as described in Section 1.5. For example, the simple code below copies the `bio` data.frame into the `tmp` data.frame.[7]

```
> tmp <- bio
```

2.2.1   Removing Individuals

There may be times when a small number of individuals (i.e., rows) must be removed from the data.frame (e.g., the individual is an outlier or data for the individual is erroneous). Rows of a data.frame are extracted by including the row position followed by a comma within [] following the data.frame name as described in Section 1.7.4.[8]

```
> bio[2,]                      # all variables for 2nd individual/row
  netID fishID  species tl   w tag scale
2    12      23 Bluegill 66 4.5    FALSE
```

An entire row of a data.frame is omitted if the row number is preceded by a negative sign, similar to what was described in Section 1.7.3 for matrices. Multiple rows are omitted by combining row numbers with `c()`.

```
> tmp <- bio[-2,]           # omit 2nd individual/row
> tmp <- bio[-c(2,7,10),] # omit 2nd, 7th, 10th individuals/rows
```

2.2.2   Filtering or Subsetting Individuals

Subsets of individuals are extracted from a data.frame with `filter()` from **dplyr** (Wickham and Francois 2015a).[9] This function requires the original data.frame as the first argument and *conditions* (Table 2.1) for filtering as additional arguments. For example, a data.frame of just Bluegill is extracted from the `bio` data.frame below.[10]

```
> bg <- filter(bio,species=="Bluegill")
> head(bg,n=3)
  netID fishID  species tl   w tag scale
1    12      16 Bluegill 61 2.9    FALSE
2    12      23 Bluegill 66 4.5    FALSE
3    12      30 Bluegill 70 5.2    FALSE
```

Multiple condition statements that are separated by commas in `filter()` (i.e., separate arguments) are treated with an "and" such that all conditions must be true for an individual to be returned.[12] For example, the code below constructs a new data.frame with only Yellow Perch that have a total length greater than or equal to 200 mm.

TABLE 2.1. Condition operators used in `filter()` and `filterD()` and their results. Note that *variable* generically represents a variable in the original data.frame and *value* is a generic value or level. Both *variable* and *value* would be replaced with specific items.

Operator	Individuals Returned from Condition with Operator
variable == value	variable is **equal** to the given value
variable ! = value	variable is **NOT** equal to the given value
variable > value	variable is **greater than** the given value
variable >= value	variable is **greater than or equal** to the given value
variable < value	variable is **less than** the given value
variable <= value	variable is **less than or equal** to the given value
*condition	condition*	**one or both conditions** are met[11]
variable %in% vector	variable is one of the values in the vector

```
> tmp <- filter(bio,species=="Yellow Perch",tl>=200)
> head(tmp,n=3)
  netID fishID    species  tl   w tag scale
1     4    113 Yellow Perch 239 150      TRUE
2     4    114 Yellow Perch 267 170      TRUE
3     4    115 Yellow Perch 262 175      TRUE
```

The `%in%` operator is used to extract individuals where a categorical variable is equal to one of several possible choices. For example, a data.frame of Black Crappie, Bluegill, and Pumpkinseed is created below.

```
> tmp <- filter(bio,species %in%
             c("Black Crappie","Bluegill","Pumpkinseed"))
```

To see if this filter was successful, one can examine the levels for the species variable with `levels()`.

```
> levels(tmp$species)
[1] "Black Crappie"   "Bluegill"        "Bluntnose Minnow"
[4] "Iowa Darter"     "Largemouth Bass" "Pumpkinseed"
[7] "Tadpole Madtom"  "Yellow Perch"
```

One notices immediately that several species are listed that should not exist in the new data.frame. A "feature" of R is to not drop the levels that existed in the original data.frame, whether or not those levels exist in the filtered data.frame. This is undesirable in most fisheries applications and is corrected by using `filterD()` from **FSA**.[13]

```
> tmp <- filterD(bio,species %in%
                 c("Black Crappie","Bluegill","Pumpkinseed"))
> levels(tmp$species)
[1] "Black Crappie" "Bluegill"      "Pumpkinseed"
```

2.2.3   Selecting Variables

Data.frames may contain more variables than the user is interested in working with. Variables are selected or removed with `select()` from **dplyr**. The first argument is the data.frame in which the variables to be selected (or removed) exist. Additional arguments identify variables to select. For example, the code below selects `species`, `tl`, and `w` from `bio`.

```
> tmp <- select(bio,species,tl,w)
> head(tmp,n=3)
    species tl   w
1 Bluegill 61 2.9
2 Bluegill 66 4.5
3 Bluegill 70 5.2
```

The same data.frame is also created by selecting a range of variable names using `:`, because these variables are contiguous in the original data.frame.

```
> tmp <- select(bio,species:w)
> head(tmp,n=3)
    species tl   w
1 Bluegill 61 2.9
2 Bluegill 66 4.5
3 Bluegill 70 5.2
```

A negative sign may be put in front of a variable name or vector of variable names to remove those variables from the data.frame. Thus, the same data.frame is constructed by eliminating the other variables.

```
> tmp <- select(bio,-c(netID,fishID,tag,scale))
> head(tmp,n=3)
    species tl   w
1 Bluegill 61 2.9
2 Bluegill 66 4.5
3 Bluegill 70 5.2
```

dplyr also provides `starts_with()`, `ends_with()`, and `contains()`, which are useful when used with `select()`. For example, the `netID` and `fishID` variables are removed in both of the following ways.

```
> tmp <- select(bio,-ends_with("ID"))
> tmp <- select(bio,-contains("ID"))
> head(tmp,n=3)
    species tl   w tag scale
1 Bluegill 61 2.9     FALSE
2 Bluegill 66 4.5     FALSE
3 Bluegill 70 5.2     FALSE
```

2.2.4 Renaming Variables

Variables are easily renamed with `rename()` from **dplyr**. The data.frame is the first argument and additional arguments define rules for renaming the variables using the format `newname=oldname`. For example, the original `tl` and `w` variables are renamed below to `length` and `weight`, respectively.

```
> tmp <- rename(bio,length=tl,weight=w)
> head(tmp,n=3)
  netID fishID  species length weight tag scale
1    12     16 Bluegill     61    2.9     FALSE
2    12     23 Bluegill     66    4.5     FALSE
3    12     30 Bluegill     70    5.2     FALSE
```

2.2.5 Creating New Variables

New variables are appended to a data.frame with `mutate()` from **dplyr**. The first argument is the original data.frame and each additional argument defines the name and rule for constructing the new variable using the format `newvar=rule`. For example, the new variables `logL` and `logW` created below contain the common logarithm[14] of the original length and weight values, respectively.

```
> tmp <- mutate(bio,logL=log10(tl),logW=log10(w))
> head(tmp,n=3)
  netID fishID  species tl   w tag scale     logL      logW
1    12     16 Bluegill 61 2.9     FALSE 1.785330 0.4623980
2    12     23 Bluegill 66 4.5     FALSE 1.819544 0.6532125
3    12     30 Bluegill 70 5.2     FALSE 1.845098 0.7160033
```

One of the strengths of `mutate()` over other similar functions is that variables created in previous arguments may be used in subsequent arguments of `mutate()`. For example, a third argument in the example above could use the `logL` or `logW` variables. A practical example of this feature is illustrated when relative weights are computed in Section 8.2.4.

2.2.6   Creating a Length Categorization Variable

Many length-based analyses in fisheries science require a variable that records the length interval to which a fish belongs (see Chapters 5 and 6). For example, the scientist may decide to categorize fish length by 25 mm wide length intervals such that a 310 mm fish would be in the 300–325 mm interval and a 473 mm fish would be in the 450–475 mm interval. Typically, length intervals for fisheries applications are *left-inclusive and right-exclusive* such that a 475 mm fish would be included in the 475–500 mm interval rather than in the 450–475 mm interval. Length intervals are usually labeled with the minimum or midpoint value of the interval. Throughout this book, intervals are labeled with the minimum value (e.g., the length interval for the 473 mm fish is 450 mm).

The `lencat()` function from **FSA** provides a flexible mechanism to create length intervals from a vector of observed lengths, which is the first argument to the function. "Rules" for creating the desired length intervals may be provided to `lencat()` in several ways. Intervals with a constant width are most common and are constructed by providing the constant width in `w=`. If only `w=` is provided, then `lencat()` will select a value for the first interval that is immediately less than the minimum observed value but consistent with `w=`. For example, if the shortest observed length is 89 mm, then the minimum value of the first interval will be 75 mm if `w=25`, 80 mm if `w=10`, and 85 mm if `w=5`. The user may also choose the starting value with `startcat=`. For example, 10 mm wide length intervals with an automatic starting category are appended to the Bluegill data.frame (created previously) in the new `lcat10` variable below.

```
> bg <- mutate(bg,lcat10=lencat(tl,w=10))
> headtail(bg)
    netID fishID   species  tl    w  tag scale lcat10
1      12     16  Bluegill  61  2.9 FALSE          60
2      12     23  Bluegill  66  4.5 FALSE          60
3      12     30  Bluegill  70  5.2 FALSE          70
218   120    807  Bluegill  69  4.0  TRUE          60
219   119    824  Bluegill 119 25.0  TRUE         110
220   122    826  Bluegill 132 33.0  TRUE         130
```

Intervals with differing widths may be created by using `c()` to create a numeric vector that contains the minimum values for the length intervals. One must make sure that the minimum and maximum values in this vector are below and above the minimum and maximum observed lengths, respectively.

The minimum and maximum values in a vector are found with `range()`, which may include `na.rm=TRUE` to remove any missing values.

```
> range(bg$tl,na.rm=TRUE)          # min & max length
[1]   27 239
```

The vector of minimum values for the intervals is then included in `breaks=` of `lencat()` to construct the desired length intervals.[15]

```
> brks <- c(0,80,150,200,250,300)  # cutoffs for intervals
> bg <- mutate(bg,lcatX=lencat(tl,breaks=brks))
> headtail(bg)
    netID fishID  species  tl    w tag scale lcat10 lcatX
1      12     16 Bluegill  61  2.9   FALSE     60     0
2      12     23 Bluegill  66  4.5   FALSE     60     0
3      12     30 Bluegill  70  5.2   FALSE     70     0
218   120    807 Bluegill  69  4.0    TRUE     60     0
219   119    824 Bluegill 119 25.0    TRUE    110    80
220   122    826 Bluegill 132 33.0    TRUE    130    80
```

By default, the length interval values in the new variable are treated as numeric values. In some instances, it is useful to treat these values as levels of a factor variable. The variable is forced to be a factor by including `as.fact=TRUE` in `lencat()`.

```
> bg <- mutate(bg,lcatX=lencat(tl,breaks=brks,as.fact=TRUE))
> levels(bg$lcatX)
[1] "0"   "80"  "150" "200" "250" "300"
```

Intervals without any fish are maintained when the interval variable is treated as a factor. If desired, use `droplevels=TRUE` in `lencat()` to drop the levels without any fish.[16]

```
> bg <- mutate(bg,lcatX=lencat(tl,breaks=brks,as.fact=TRUE,
                               droplevels=TRUE))
> levels(bg$lcatX)
[1] "0"   "80"  "150" "200"
```

Finally, if the vector of interval values is named (see Section 1.7.1), then those names may be used in place of the values if `use.names=TRUE` is given to `lencat()`. In this case, the result is considered to be a factor and `droplevels=` may be used to drop the unused levels from consideration.

```
> ( brks <- c(Z=0,S=80,Q=150,P=200,M=250,T=300) )
  Z   S   Q   P   M   T
  0  80 150 200 250 300
> bg <- mutate(bg,lcatX=lencat(tl,breaks=brks,use.names=TRUE))
> levels(bg$lcatX)
[1] "Z" "S" "Q" "P" "M" "T"
```

2.2.7   Recoding Variables

Situations may occur where the values of a variable may need to be changed. Recoding values is efficiently accomplished with mapvalues() from **plyr** (Wickham 2011),[17] where the first argument is a vector that contains the values to be recoded. The original values to be recoded are entered into a vector that is given to from=. The corresponding new values are entered into a vector (the same size as the vector of original values) and given to to=.

For example, the one-letter names in the lcatX variable added to the bg data.frame above are recoded below to more descriptive labels. The codes in the original lcatX variable and the corresponding codes for the new variable are first entered into separate vectors, making sure that the codes matched one-to-one between the two vectors.

```
> old <- c("Z","S","Q","P","M","T")
> new <- c("Sub-Stock","Stock","Quality","Preferred",
           "Memorable","Trophy")
```

Finally, mapvalues() is used within mutate() to create the new variable (lcatY) with new codes. For succinctness in this illustration, several variables are first removed from the data.frame.

```
> bg <- select(bg,-c(netID,fishID,lcat10))
> bg <- mutate(bg,lcatY=mapvalues(lcatX,from=old,to=new))
> headtail(bg)
      species  tl    w tag scale lcatX      lcatY
1    Bluegill  61  2.9     FALSE     Z Sub-Stock
2    Bluegill  66  4.5     FALSE     Z Sub-Stock
3    Bluegill  70  5.2     FALSE     Z Sub-Stock
218  Bluegill  69  4.0      TRUE     Z Sub-Stock
219  Bluegill 119 25.0      TRUE     S     Stock
220  Bluegill 132 33.0      TRUE     S     Stock
```

The from= and to= vectors may have fewer than all possible values, in which case only the values given in from= will be changed and all other values in the original variable will be retained. Additionally, two from= values may be mapped to the same to= value to effectively combine values. For example, the two vectors below could be used to recode the `lcatX` variable as above, with the exception that all of the original `"P"`, `"M"`, and `"T"` values are mapped to `"Preferred"` in the new variable.

```
> old <- c("Z","S","Q","P","M","T")
> new <- c("Sub-Stock","Stock","Quality","Preferred",
           "Preferred","Preferred")
```

2.2.8 Sorting

All rows in a data.frame may be sorted by values of one or more variables with `arrange()` from **dplyr**. The original data.frame is the first argument and is followed by one or more arguments that contain variables which describe how the rows should be sorted. Values are sorted in ascending order unless the variable is wrapped in `desc()`, in which case it will be sorted in descending order. The rows are sorted by the first variable provided (i.e., second argument) and then within common values of the first variable by the second variable provided, and so on. The example below sorts the `bio` data.frame by net identification number, then species, and then by descending total length within each net ID and species.

```
> tmp <- arrange(bio,netID,species,desc(tl))
> headtail(tmp)
    netID fishID        species    tl   w    tag scale
1       1    103  Black Crappie   284  NA  g0998  TRUE
2       1     93 Largemouth Bass  386  NA  o0447  TRUE
3       1     98 Largemouth Bass  371  NA  g1000  TRUE
674   205    522 Bluntnose Minnow  61 2.1        FALSE
675   205    521    Iowa Darter    53 1.1        FALSE
676   206    501       Bluegill    38 0.7        FALSE
```

2.2.9 Combined Manipulations with Piping

A data.frame (`df`) is "piped" into the first argument of a function (`FUN()`) with the `%>%` operator from **dplyr**[18] using the form `df %>% FUN()`. For example, the last example of `arrange()` above is also accomplished with `%>%` below because `bio` is piped into the first argument and `netID` becomes the second argument of `arrange()`.

```
> tmp <- bio %>% arrange(netID,species,desc(tl))
```

The real advantage of `%>%`, however, is that it allows one to chain together multiple data manipulation steps. When a function that returns a data.frame is before %>%, then its result is piped into the first argument of the function that follows %>%. Thus, the data.frame returned by one function is immediately sent to the succeeding function, and so on.

When reading code, replace %>% with "then." For example, the code below creates a data.frame object called bg that is created by filtering Bluegill from bio, *THEN* deleting the tag and scale variables, *THEN* adding the common logarithms of total length and weight and 10 mm length categories, and *THEN* sorting the data.frame by netID and total length.

```
> bg <- bio %>%
    filter(species=="Bluegill") %>%
    select(-tag,-scale) %>%
    mutate(logL=log10(tl),logW=log10(w),
           lcat=lencat(tl,w=10)) %>%
    arrange(netID,tl)

> head(bg,n=3)
  netID fishID  species  tl  w      logL      logW lcat
1     4    134 Bluegill  99 15 1.995635 1.176091   90
2     4    125 Bluegill 135 35 2.130334 1.544068  130
3     4    118 Bluegill 150 60 2.176091 1.778151  150
```

The %<>% operator from **magrittr** (Bache and Wickham 2014) is used when a data.frame should be piped into the first argument of a function **AND** the result of the function should be assigned to that same data.frame. For example, the code below pipes the bg data.frame into the first argument of filter() and the result from filter() is assigned back to bg. Thus, this code modifies the bg data.frame to include only fish greater than 100 mm.

```
> bg %<>% filter(tl>100)
> head(bg,n=3)
  netID fishID  species  tl  w      logL      logW lcat
1     4    125 Bluegill 135 35 2.130334 1.544068  130
2     4    118 Bluegill 150 60 2.176091 1.778151  150
3     4    121 Bluegill 152 67 2.181844 1.826075  150
```

The same data.frame would have been constructed with the following code using only %>% from dplyr.

```
> bg <- bg %>% filter(tl>100)
```

2.3 Joining Data.Frames

2.3.1 Combining Similar Data.Frames

Data.frames with identical structures (number AND names of variables) may be vertically combined to form one larger data.frame with `rbind()`[19] (Figure 2.2). The `rbind()` function takes two or more data.frames as arguments and *row-binds* these together such that each subsequent data.frame forms rows below the previous data.frame. If the data.frames to be given to `rbind()` do not have the same variables, then "extra" variables need to be deleted from one data.frame or added (and filled with `NA` values) to the other data.frame. Furthermore, if the variable names differ, then they must be modified to be the same for both data.frames.

Source A

net	lake	code
1	A	BG
1	A	LMB

Source B

net	lake	code
2	A	BG
2	A	LMB
2	A	SMB

Row Bound

net	lake	code
1	A	BG
1	A	LMB
2	A	BG
2	A	LMB
2	A	SMB

FIGURE 2.2. Demonstration of row-binding two source data.frames.

As an example, *inchBio07.csv* and *inchBio08.csv* (loaded below into `bio07` and `bio08`, respectively) contain the same information as `bio`, but separated for 2007 and 2008, respectively. The structures for these two data.frames show that they have the same number of variables and the variables seem to represent the same information. However, because different names are used for the first variable in these two data.frames, they cannot be immediately combined with `rbind()`.

```
> bio07 <- read.csv("inchBio07.csv")
> str(bio07)
'data.frame': 349 obs. of  7 variables:
 $ netID  : int  1 1 1 1 1 1 1 1 1 1 ...
 $ fishID : int  90 91 92 93 94 95 96 97 98 99 ...
 $ species: Factor w/ 7 levels "Black Crappie",..: 5 5 5 5 5 5..
 $ tl     : int  356 328 305 386 310 262 305 315 371 338 ...
 $ w      : num  NA NA NA NA NA NA NA NA NA NA ...
 $ tag    : Factor w/ 97 levels "","g0985","g0986",..: 93 17 1..
 $ scale  : logi  TRUE TRUE TRUE TRUE TRUE TRUE ...
> bio08 <- read.csv("inchBio08.csv")
> str(bio08)
'data.frame': 327 obs. of  7 variables:
 $ net    : int  101 101 101 101 101 101 101 101 101 101 ...
 $ fishID : int  532 533 534 535 536 537 538 539 540 541 ...
 $ species: Factor w/ 8 levels "Black Crappie",..: 1 2 1 1 5 1..
 $ tl     : int  320 213 305 307 318 287 216 216 231 193 ...
 $ w      : num  508 190 443 440 407 379 198 210 258 138 ...
 $ tag    : Factor w/ 102 levels "","1014","1015",..: 2 1 4 5 ..
 $ scale  : logi  TRUE TRUE TRUE TRUE TRUE TRUE ...
```

The net variable in bio08 is changed below to netID with rename() (as in Section 2.2.4) to match bio07.

```
> bio08 %<>% rename(netID=net)
```

Now that these two data.frames have the **exact** same number and names of variables, they may be vertically combined with rbind() to form a new data.frame.

```
> tmp <- rbind(bio07,bio08)
> str(tmp)
'data.frame': 676 obs. of  7 variables:
 $ netID  : int  1 1 1 1 1 1 1 1 1 1 ...
 $ fishID : int  90 91 92 93 94 95 96 97 98 99 ...
 $ species: Factor w/ 8 levels "Black Crappie",..: 5 5 5 5 5 5..
 $ tl     : int  356 328 305 386 310 262 305 315 371 338 ...
 $ w      : num  NA NA NA NA NA NA NA NA NA NA ...
 $ tag    : Factor w/ 193 levels "","g0985","g0986",..: 93 17 ..
 $ scale  : logi  TRUE TRUE TRUE TRUE TRUE TRUE ...
```

2.3.2 Combining Data.Frames by a Key Relation

Two data.frames may be linked by a *key* variable. For example, one data.frame may contain all information about each sampling event and a second data.frame may contain information about each fish captured in each sampling event. Records in these data.frames may be linked through a variable in each data.frame that contains a unique code for each sampling event. This is the same concept used to link multiple tables in simple relational databases. The variable that links the two data.frames is called the "key."

Data.frames may be merged or joined such that information in one data.frame is added to the information in another data.frame based on the value of the key variable. There are several types of data.frame "joins" (deVries and Meys 2012; Figure 2.3).

- An *inner* or *natural* join returns only those rows for which values of the key are present in **both** data.frames.
- A *left* join returns all rows from the first data.frame. If a key value is present in the first but not the second data.frame, then an NA is returned for the variables from the second data.frame.
- An *outer* join returns all rows from each data.frame. If a key value is present in one but not the other data.frame, then an NA is returned for the variables of the other data.frame.

Source A

net	lake
1	A
2	A
3	B

Source B

net	code
1	BG
1	LMB
2	BG
4	BG

Inner Join

net	lake	code
1	A	BG
1	A	LMB
2	A	BG

Left Join

net	lake	code
1	A	BG
1	A	LMB
2	A	BG
3	B	NA

Outer Join

net	lake	code
1	A	BG
1	A	LMB
2	A	BG
3	B	NA
4	NA	BG

FIGURE 2.3. Demonstration of an inner, left, and outer join from two source data.frames with the key variable **net**. Concept modified from RStudio (2015a).

These three joins are accomplished with inner_join(), left_join(),[20] and outer_join() from **dplyr**.[21] The first two arguments to each of these functions are the data.frames to be joined. The key value variable is given in by=.

The `bio` data.frame examined in previous sections is an example of a data.frame that contains specific information about each fish sampled, including a unique code for the sampling event in which the fish was captured. Information specific to each sampling event – unique net identification code, type of gear used, deployment date (year, month, day), temperature, and effort expended – is recorded in *inchGear.csv* (loaded into `gear` below). Information in these two data.frames are linked through the `netID` key value variable.

```
> gear <- read.csv("inchGear.csv")
> head(gear,n=3)
  netID netType year month day temp unitsOfEffort
1     1 angling 2007   May  13   NA          5.50
2     2 angling 2007   May  13   NA          8.25
3     3 angling 2007   May  14   NA          0.50

> head(bio,n=3)          # reminder of what bio looks like
  netID fishID  species tl   w tag scale
1    12     16 Bluegill 61 2.9     FALSE
2    12     23 Bluegill 66 4.5     FALSE
3    12     30 Bluegill 70 5.2     FALSE
```

An inner join would return only those rows from `gear` and `bio` where a value for the `netID` variable existed in both data.frames. This would result in a data.frame that did not list gears for which no fish data existed. This result may be problematic if the ultimate goal is to count the number of fish for the purpose of computing catch per unit effort per `netID`. In this case, a left join would be more appropriate as it will ensure that the information from all sampling events in `gear` is included in the final joined data.frame, regardless of whether or not fish from that sampling event were recorded in the `bio` data.frame. Code for a left join is illustrated below, with some variables eliminated and one renamed so that the result could be more easily seen. Note that the last several rows of the resulting data.frame contain `NA` values for the variables after `fishID` because no fish were captured in those sampling efforts.

```
> bio2 <- left_join(gear,bio,by="netID") %>%
    rename(effort=unitsOfEffort) %>%
    select(netID,netType,year,effort,fishID,species,tl)
> headtail(bio2)
    netID netType year effort fishID         species  tl
1       1 angling 2007    5.5     90 Largemouth Bass 356
2       1 angling 2007    5.5     91 Largemouth Bass 328
3       1 angling 2007    5.5     92 Largemouth Bass 305
717   221   seine 2008    1.0     NA            <NA>  NA
718   222   seine 2008    1.0     NA            <NA>  NA
719   223   seine 2008    1.0     NA            <NA>  NA
```

As a second example, one may want to join the biological data in `bio2` with aging data, which may include measurements made on a calcified structure (e.g., scales) from each fish. For example, *inchAge.csv* (loaded below into `age`) contains a unique code for each fish, which scale was assessed (more than one scale was used for most fish), assigned age, length from the focus to the margin of the scale at the time of capture (`radCap`), and lengths from the focus to each annulus (`radX` with X replaced by the annulus number). The information in `bio2` and `age` are linked with the `fishID` key.

```
> age <- read.csv("inchAge.csv")
> head(age,n=4)
  fishID time ageCap radCap   rad1   rad2   rad3   rad4 rad5
1     90    1      4 4.3868 1.9986 3.1746 3.7302 4.3868   NA
2     90    2      4 4.5599 2.0779 3.1457 3.9249 4.5599   NA
3     91    1      4 4.0692 1.5007 2.2799 3.3405 4.0692   NA
4     91    2      4 3.5858 1.3564 2.0490 3.0086 3.5858   NA
  rad6 rad7
1   NA   NA
2   NA   NA
3   NA   NA
4   NA   NA
```

The biological information for each fish is joined below with the age information for each scale. An inner join is used here so that the final data.frame contains information only for those fish for which an age was assigned **AND** biological information was available. Several variables were subsequently removed to save space, as this data.frame will be used in the next section.

```
> age2 <- merge(bio2,age) %>%
    select(-c(netID,netType,effort,ageCap))
> head(age2,n=4)
  fishID year         species  tl time radCap   rad1   rad2
1     90 2007 Largemouth Bass 356    1 4.3868 1.9986 3.1746
2     90 2007 Largemouth Bass 356    2 4.5599 2.0779 3.1457
3     91 2007 Largemouth Bass 328    2 3.5858 1.3564 2.0490
4     91 2007 Largemouth Bass 328    1 4.0692 1.5007 2.2799
    rad3   rad4 rad5 rad6 rad7
1 3.7302 4.3868   NA   NA   NA
2 3.9249 4.5599   NA   NA   NA
3 3.0086 3.5858   NA   NA   NA
4 3.3405 4.0692   NA   NA   NA
```

2.4 Rearranging Data.Frames

The age2 data.frame from the previous section may be arranged in two different ways. If the fish is the individual (i.e., unit of observation), then age2 is in *long format*, where each row contains information about only one individual. In other words, each row contains data about one, and only one, fish. However, if an annulus is the individual, then age2 is in *wide format*, where the results for many individuals appear in one row. In a wide format, some variables may correspond to multiple individuals (e.g., the species is relevant for each annulus), but values of the primary response variable (radial measurement to the annulus) that correspond to different individuals (e.g., annuli) appear in the same row.

A fisheries scientist may need to convert between formats depending on the type of analysis. For example, the data must be in long format with the fish as the individual (as with age2) to examine the relationship between the size of the scale and the size of the fish. However, the data must be in wide format with the annulus as the individual to back-calculate fish length at a previous age.[22]

Data are converted from wide to long format with **gather()** from **tidyr** (Wickham 2014c), using the wide format data.frame as the first argument. The second argument is a name for the new variable that will identify the individual in the long format data.frame to be created (created from the corresponding column names in the wide format). The third argument is a name for the new variable that will contain the observations for an individual in the long format data.frame to be created (which are the values in the columns in the wide format data.frame). The fourth argument is the variable names in the original wide format data.frame that contain the values of the primary response variable. The variables in the fourth argument may be selected with the : operator if the variables are contiguous (as illustrated below), by combining the variable names with c(), or by eliminating variables with the negative sign. An illustration of this process is shown in Figure 2.4.

Data are converted from long to wide format with **spread()** from **tidyr**, using the long format data.frame as the first argument. The second argument is the name of the variable in the long format that identifies the individual (will form the column names in the wide format data.frame to be created). The third argument is the name of the variable in the long format data.frame that contains the observations for an individual (will fill the columns in the wide format data.frame to be created). An illustration of this process is shown in Figure 2.5.

The data in the wide format age2 data.frame are converted to a new long format data.frame with the radii "names" in the new variable age and the measurements in the new variable rad with the code below. Note that each fish appears in as many rows as there are annuli (seven in this example).

Original Wide Format

fish	code	rad1	rad2
1	BG	1.3	1.1
2	LMB	1.6	1.4

New Long Format

fish	code	num	meas
1	BG	rad1	1.3
1	BG	rad2	1.1
2	LMB	rad1	1.6
2	LMB	rad2	1.4

FIGURE 2.4. Demonstration of converting from a wide to a long format data.frame with `gather(wide,num,meas,3:4)`. Concept modified from RStudio (2015a).

Original Long Format

fish	code	num	meas
1	BG	rad1	1.3
1	BG	rad2	1.1
2	LMB	rad1	1.6
2	LMB	rad2	1.4

New Wide Format

fish	code	rad1	rad2
1	BG	1.3	1.1
2	LMB	1.6	1.4

FIGURE 2.5. Demonstration of converting from a long to a wide format data.frame with `spread(long,num,meas)`. Concept modified from RStudio (2015a).

```
> age2L <- gather(age2,age,rad,rad1:rad7)
> headtail(age2L)
     fishID year         species  tl time radCap  age    rad
1        90 2007 Largemouth Bass 356    1 4.3868 rad1 1.9986
2        90 2007 Largemouth Bass 356    2 4.5599 rad1 2.0779
3        91 2007 Largemouth Bass 328    2 3.5858 rad1 1.3564
3029    913 2008 Largemouth Bass 251    2 3.2637 rad7     NA
3030    914 2008 Largemouth Bass 378    1 4.8920 rad7     NA
3031    914 2008 Largemouth Bass 378    2 5.1009 rad7     NA
```

There are at least two issues with the resulting `age2L` data.frame. First, the `age` variable data are "rad1", "rad2", and so on rather than 1, 2, and so on. Second, several rows contain no data (i.e., `NA`) for `rad` because not all annuli exist given the observed age of the fish (i.e., there cannot be a sixth annulus for a five-year-old fish). These issues result largely from using this type of data (i.e., radial measurements) as an example rather than a feature of `gather()`. Thus, methods to address these two issues are illustrated in the online supplement that demonstrates back-calculation of previous length.

As an example, these data are converted back to wide format as follows.

```
> age2W <- spread(age2L,age,rad)
> head(age2W,n=3)
  fishID year        species   tl time radCap   rad1    rad2
1     90 2007 Largemouth Bass 356    1 4.3868 1.9986 3.1746
2     90 2007 Largemouth Bass 356    2 4.5599 2.0779 3.1457
3     91 2007 Largemouth Bass 328    1 4.0692 1.5007 2.2799
    rad3   rad4 rad5 rad6 rad7
1 3.7302 4.3868   NA   NA   NA
2 3.9249 4.5599   NA   NA   NA
3 3.3405 4.0692   NA   NA   NA
```

The `gather()` and `spread()` functions may be chained together with the pipe operator (`%>%`) because their first arguments and results are data.frames.

2.5   New Data.frame from Aggregation

New data.frames may also be created from calculations on existing data.frames. Most often the new data.frames will consist of data computed for each group in the original data.frame. The process of computing a summary for groups in a data.frame and returning the results for each group in a new data.frame is called *aggregation*. Aggregation is generally a three-step process of splitting the original data into groups, applying a summary function to the data in each group, and then combining the results from each group together to form the new data.frame (Wickham 2011; Figure 2.6).

A variable that defines the groups is declared with `group_by()` from **dplyr**. The first argument to `group_by()` is a data.frame and subsequent arguments are variables that define groups in the data.frame. For example, a grouping by `netID` and `species` within `netID` is added to the `bio` data.frame below.

```
> tmp <- group_by(bio,netID,species)
```

Additional classes are added to a data.frame by `group_by()`.

```
> class(tmp)
[1] "grouped_df" "tbl_df"     "tbl"        "data.frame"
```

Original	
sex	len
F	17
F	19
M	13
M	15
M	20
J	13

Split	
sex	len
F	17
F	19
M	13
M	15
M	20
J	13

Combined	
sex	AVG
F	18
M	16
J	13

FIGURE 2.6. Demonstration of using the "split-apply-combine" strategy to create a new data.frame by computing the average length for each sex in the original data.frame. Concept modified from RStudio (2015a).

These new classes hold the grouping variable information and result in a different display of the data.frame (i.e., only 10 rows and as many columns as can fit the width of the device are shown when the default settings are used). These changes do not effect what has been illustrated previously, except that the display of the data.frame will appear different.

```
> tmp
Source: local data frame [676 x 7]
Groups: netID, species [121]

   netID fishID species  tl    w    tag scale
   (int)  (int)   (fctr) (int) (dbl) (fctr) (lgl)
1     12     16 Bluegill  61   2.9        FALSE
2     12     23 Bluegill  66   4.5        FALSE
3     12     30 Bluegill  70   5.2        FALSE
4     12     44 Bluegill  38   0.5        FALSE
5     12     50 Bluegill  42   1.0        FALSE
6     12     65 Bluegill  54   2.1        FALSE
7     12     66 Bluegill  27    NA        FALSE
8     13     68 Bluegill  36   0.5        FALSE
9     13     69 Bluegill  59   2.0        FALSE
10    13     70 Bluegill  39   0.5        FALSE
..    ...    ...     ...  ...   ...   ...   ...
```

Summary functions are applied to the groups and the results combined together with summarize() from **dplyr**. The first argument to summarize() is the data.frame. Subsequent arguments define a type of summary by using newvar=FUN, where newvar is the name of the new variable in the resulting data.frame and FUN is the name of a function that will produce the summary. The function in FUN must take a vector and return a single value. Typical functions include mean() for sample means, sd() for sample standard deviations, min() and max() for the minimum and maximum values, quantile() for quantiles, n() from **dplyr** for sample sizes, and validn() from FSA for sample sizes excluding missing values (that are coded with NAs).

One level of the grouping structure (i.e., the inner-most level) is removed from the data.frame with each use of summarize(). For example, if summarize() was applied to the bio data.frame, then the grouping based on species would be removed. However, the grouping based on netID would remain. This is a valuable feature because it allows the user to summarize data.frames in a hierarchical manner. However, if one is done summarizing a data.frame, then it is best to submit that data.frame to ungroup() from **dplyr** or as.data.frame() to remove any remaining grouping structures. In this book, as.data.frame() is used as it also removes the tbl.df class so that the whole data.frame will be printed when asked.

Most aggregations can be accomplished with the "split-apply-combine" concept (Wickham 2011) described above. Three examples are used below to illustrate the range of tasks that can be accomplished with this concept.

Multiple measurements may be made on the same individuals in an effort to either control or understand measurement variability. For example, measurements were made on multiple scales from the same fish in the age data.frame used previously. The fisheries scientists may want to average the measurements across the multiple scales to create measurements for an "average" scale. To facilitate these calculations, the data in the original age data.frame are grouped by the fish identification variable. The mean for age-at-capture, scale radial measurement-at-capture, and each radial measurement is then computed with the final result assigned to a new data.frame.[23]

```
> tmp <- age %>% group_by(fishID) %>%
    summarize(aageCap=mean(ageCap),aradCap=mean(radCap),
              arad1=mean(rad1),arad2=mean(rad2),arad3=mean(rad3),
              arad4=mean(rad4),arad5=mean(rad5),arad6=mean(rad6),
              arad7=mean(rad7)) %>% as.data.frame()
> head(tmp,n=2)
  fishID aageCap aradCap   arad1   arad2   arad3   arad4 arad5
1     90       4 4.47335 2.03825 3.16015 3.82755 4.47335    NA
2     91       4 3.82750 1.42855 2.16445 3.17455 3.82750    NA
  arad6 arad7
1    NA    NA
2    NA    NA
```

If a data.frame contains biological information for every individual captured, then the individuals could be counted for each sampling event and species as a record of catch. For example, the code below groups the bio data.frame from above by netID and then species, then counts the number of records (i.e., using n()) within each grouping, and assigns the result to a data.frame. This data.frame could then be joined, via the key value variable netID, with a data.frame that contains sampling event information for further analyses (e.g., compare catch of a species among years, gears, or habitats).

```
> tmp <- bio %>% group_by(netID,species) %>%
    summarize(catch=n()) %>%
    as.data.frame()
> head(tmp,n=3)
  netID          species catch
1     1    Black Crappie     1
2     1 Largemouth Bass    13
3     2 Largemouth Bass     9
```

Finally, the example below shows how to use group_by() and summarize() to compute summaries of lengths for each species in each capture year in the bio2 data.frame created previously. Note the use of na.rm=TRUE to remove missing values from the vector so that the summary is properly computed from the existing data. Also note the use of round() to round the result to the number of decimal places in its second argument.

```
> tmp <- bio2 %>% group_by(year,species) %>%
    summarize(n=n(),val.n=validn(tl),
            mean=round(mean(tl,na.rm=TRUE),1),
            sd=round(sd(tl,na.rm=TRUE),2),
            min=min(tl,na.rm=TRUE),
            mdn=quantile(tl,0.5,na.rm=TRUE),
            max=max(tl,na.rm=TRUE)) %>%
    as.data.frame()
> head(tmp,n=3)
  year          species   n val.n  mean    sd min mdn max
1 2007    Black Crappie  15    15 281.5 37.43 152 290 305
2 2007         Bluegill 118   118 138.5 63.22  27 152 234
3 2007 Bluntnose Minnow  51    51  62.2  6.23  51  61  78
```

2.5.1 Summarizing Catch per Unit Effort

Fisheries scientists use catch per unit effort (CPE or CPUE) to standardize the number of fish caught by the amount of effort used to catch those fish. For example, CPE may represent the number of fish captured per one fyke net set for one night, per one hour of shoreline electrofishing, per 1000 meters of gillnet, or per 100 hooks. CPE is the ratio of the number of fish caught to the effort expended. The effort should be converted to the desired units prior to computing this ratio. For example, if 16 Largemouth Bass were captured in 10 minutes of electrofishing, then the CPE expressed in fish per hour would be 96 $(= \frac{16}{10/60})$.

Summarizing CPE by species requires many of the data manipulation tasks described in this chapter and, thus, provides a useful case study. In the example below, one method of summarizing CPE by species is demonstrated using data on the sample (e.g., date, effort) and fish caught in the sample (e.g., species, length) from a Kansas lake.

Data regarding each sampling event are loaded into the `gear` object below. Note that `UID` is the key value variable that identifies the unique sampling event. Effort was recorded in minutes of electrofishing and was immediately converted to hours below.

```
> gear <- read.csv("BGHRsample.csv") %>%
    mutate(effort=effort/60)
> head(gear,n=3)
  UID      date loc    effort
1   1 5/15/2014 196 0.1666667
2   2 5/15/2014 200 0.1666667
3   3 5/14/2014 201 0.1666667
```

The fish specific data are loaded into the `fish` object below. Species names were originally recorded as a numeric code so `mapvalues()` was used to convert those codes to descriptive names. Individuals that were measured have a value in `length` and `count` is one. Some fish were not measured for length. In these cases, the species is listed, no length is given, and the number of captured fish that were not measured is recorded in `count`. Also note that `UID` is the key value variable that connects these data to the sampling event data in `gear`.

```
> fish <- read.csv("BGHRfish.csv") %>%
    mutate(spec=mapvalues(specCode,from=c(116,118,122),
           to=c("Smallmouth Bass","Largemouth Bass","Bluegill")))
> headtail(fish)
    UID fishID specCode length weight count            spec
1     1      1      116    100     NA     1 Smallmouth Bass
2     1      2      116    100     NA     1 Smallmouth Bass
3     1      3      118    298    300     1 Largemouth Bass
264  20     19      122    166    100     1        Bluegill
265  20     20      122    168    110     1        Bluegill
266  20     21      122    171    120     1        Bluegill
```

The numbers of each species caught in each sampling event are summarized from the `fish` data.frame by first grouping the data by fish species within each unique sampling event and then summing the values in `count`.

```
> catch <- fish %>% group_by(UID,spec) %>%
    summarize(caught=sum(count)) %>%
    as.data.frame()
```

The catch data.frame is then left-joined with the sampling event data.frame so that the number captured is connected with the amount of effort for each sampling event.

```
> catch <- left_join(gear,catch,by="UID")
> headtail(catch)
   UID    date loc    effort           spec caught
1    1 5/15/2014 196 0.1666667       Bluegill     18
2    1 5/15/2014 196 0.1666667 Largemouth Bass      9
3    1 5/15/2014 196 0.1666667 Smallmouth Bass      2
39  19 5/15/2014 270 0.1666667 Largemouth Bass     14
40  20 5/15/2014 281 0.1666667       Bluegill     15
41  20 5/15/2014 281 0.1666667 Largemouth Bass     19
```

However, one should examine the resulting data.frame very carefully. For example, in the snippet above, note that the sampling event labeled with a UID of 20 does not show any Smallmouth Bass. In fact, none of the species that had a "catch" of zero in a sampling event are shown in this data.frame. This is problematic because the mean CPE computed on these data would be too large because no zeroes would be included in the calculation of the mean.

Zeroes are efficiently added to the data.frame with `addZeroCatch()` from **FSA**. This function requires the names (in quotes) of the sampling event key value variable as the first argument and the variable that contains the species names as the second argument. Variables to receive the zeroes are given in `zerovar=`. With this information, `addZeroCatch()` identifies all sampling event and species combinations that do not exist in the data.frame, creates new entries with zeroes in the variables given in `zerovar=` for those combinations (and all other related variables not given in `zerovar=`), and appends those new entries to the original data.frame. Thus, the returned data.frame contains data for all sampling event and species combinations.

```
> catch %<>% addZeroCatch("UID","spec",zerovar="caught") %>%
    arrange(UID,spec)
```

A `cpe` variable that contains the catch of each species per **hour** of electrofishing can now be appropriately appended to the data.frame.

```
> catch %<>% mutate(cpe.hr=caught/effort)
> headtail(catch)
   UID      date loc   effort            spec caught cpe.hr
1    1 5/15/2014 196 0.1666667        Bluegill     18    108
2    1 5/15/2014 196 0.1666667 Largemouth Bass      9     54
3    1 5/15/2014 196 0.1666667 Smallmouth Bass      2     12
58  20 5/15/2014 281 0.1666667        Bluegill     15     90
59  20 5/15/2014 281 0.1666667 Largemouth Bass     19    114
60  20 5/15/2014 281 0.1666667 Smallmouth Bass      0      0
```

Take special note of the zero catch and CPE for Smallmouth Bass in the last sample shown above.

These data may be summarized to find the total number of sampling events, the total number of fish, and the mean ($\bar{x}$), standard deviation, standard error (SE), and relative standard error ($RSE = 100 * \frac{SE}{\bar{x}}$; Koch et al. 2014) for each species.

```
> cpeSum <- catch %>% group_by(spec) %>%
    summarize(samples=n(),fish=sum(caught),
           mean=mean(cpe.hr),sd=sd(cpe.hr),
           se=sd/sqrt(samples),RSE=se/mean*100) %>%
    as.data.frame()
> cpeSum
             spec samples fish mean        sd       se      RSE
1        Bluegill      20  158 47.4 42.910433 9.595065 20.24275
2 Largemouth Bass      20  128 38.4 33.058799 7.392172 19.25045
3 Smallmouth Bass      20   10  3.0  6.882472 1.538968 51.29892
```

2.6   Exporting Data.Frames to External Data Files

Code used to analyze data should be saved in a script as described in Section 1.10. However, some analyses may produce a data.frame that one may want to export to an external file for use in another script or with other software. The results of an object are exported to a CSV file with `write.csv()`.[24] The name of the object in R is the first argument to `write.csv()` and the name of the external CSV file, in quotes, is the second argument. Additionally, it is common to set `quote=FALSE` so that the results are not contained in quotes and `row.names=FALSE` so that row names are not included in the file. The code below writes the `tmp` data.frame from the previous section to the external `"output.csv"` file in the current working directory.

```
> write.csv(tmp,"output.csv",quote=FALSE,row.names=FALSE)
```

2.7   Further Considerations

Much of what is illustrated in Sections 2.2-2.5 can be accomplished with functions in base R, rather than using the **dplyr** and **tidyr** packages. The **dplyr** and **tidyr** options are illustrated here because they provide efficient (in terms of keystrokes) and very readable code. Other **dplyr** and **tidyr** functions that are not illustrated in this chapter also provide very usable functionality to the fisheries scientist (but not at the introductory level of this book). The speed of the **dplyr** and **tidyr** functions described here are sometimes faster and sometimes slower than the alternatives in base R.

One problem with using the functions described in this chapter is that functions in other packages have the same name. For example, both **dplyr** and **plyr** have `rename()` and **dplyr** and base R have `filter()`. For two functions with the same name, R should use the function from the package that was most recently loaded. However, to ensure that the function from a particular package is used, the user can explicitly state the package to be used with `package::function()`. For example, `dplyr::rename()` will ensure that `rename()` from **dplyr** is used. One does not need to include the package name in the function call if two packages with functions of the same name are not loaded at the same time.

Many of the data manipulations described in this chapter can also be accomplished with functions from **data.table** (Dowle et al. 2014)

Finally, **readr** (Wickham and Francois 2015b) and **rio** were released as this book was being released. Both packages hold much promise for simplifying the process of reading data into R.

Notes

[1] If the working directory is many directories "deep," then, if using RStudio, it may be easier to set the working directory with one of the methods under the "Session" menu and "Set Working Directory" submenu. While this use of menus is convenient, the resulting `setwd()` code should be pasted into your script so that future uses of the script will not require having to use the menu options (i.e., the script should be self-contained). Additionally, use of "projects" in RStudio largely obviates the need to set the working directory.

[2] A single backslash as used by the Windows operating system will not work.

[3] The relatively new **readxl** package can be used to read data from Excel without installing other software.

[4] Here, "text files" refers to the very simple text or ASCII files discussed in Section 2.1.3. However, it should be noted that the common "comma separated values" file type is a form of a text file.

[5] The `levels()` functions expects a vector; thus, the variable must be given in `df$var` format as described in Section 1.7.4.

[6] If using RStudio, the name of the data.frame object can be selected from the "Envi-

ronment" pane (usually the upper-right pane) to open the object in the main RStudio pane.

[7]Throughout this book, the `tmp` object is used when an object is created as part of an example but will not be used further in the chapter (i.e., it is temporary).

[8]Recall that the comma inside the square brackets separates the row and column positions in the data.frame. If no position is given for the columns, then all columns are returned.

[9]Several other R packages have a function named `filter()`. If these other packages are loaded after **dplyr** then `filter()` from **dplyr** will not be used and an error will likely follow. To ensure that `filter()` from **dplyr** is used, then use `dplyr::filter()`.

[10]This is shown in Table 2.1, but make special note of the use of the "double equals" (`==`) when forming the "is equal to" condition.

[11]The "or" operator is a "vertical line" which is typed with the shift-backslash key on most keyboards.

[12]Multiple conditions are explicitly joined with an "and" by placing the `&` operator between conditions. For example, `filter(bio,species=="Yellow Perch" & tl>=200)` is the same as `filter(bio,species=="Yellow Perch",tl>=200)`.

[13]Unused levels in a factor variable can also be dropped by including the factor vector in `droplevels()`. The `filterD()` function is simply `filter()` followed by `droplevels()`.

[14]Natural logarithms are used in many fisheries applications. Natural logarithms are used instead of common logarithms by replacing `log10()` with `log()`.

[15]If `breaks=` is used, then both `w=` and `startcat=` are ignored.

[16]The intervals with no fish are automatically dropped when the intervals are numeric.

[17]The `mapvalues()` function is originally from the **plyr** package, but is exported by **FSA** so that **plyr** does not have to be explicitly installed. This reduces conflicts with functions of the same name in **dplyr** and **plyr**.

[18]Actually, `%>%` is originally from **magrittr** but it is exported and, thus, available to the user from **dplyr**.

[19]This same task is also accomplished with `bind_rows()` from **dplyr**.

[20]There is a `right_join()` in **dplyr**. However, a right join can also be accomplished by reversing the order of the data.frames in `left_join`.

[21]These joins, and others, can also be accomplished with `merge()` in base R.

[22]A more thorough example of back-calculating previous length is available in the online chapter supplement.

[23]The data in `age` have been prescreened such that the same age was determined for both scales from the same fish. If this were not the case, then the means shown here might not make biological sense and would need to be screened in R.

[24]Data may also be exported using `export()` from **rio**.

3

Plotting Fundamentals

Fisheries scientists often plot their data or summary statistics to gain further understanding or to communicate their findings. R provides at least three graphics systems: *base*, *ggplot* through **ggplot2** (Wickham 2009; Chang 2012), and *trellis* or *lattice* through **lattice** (Sarkar 2008). This chapter introduces the base graphics system to construct five plots commonly used by fisheries scientists and in this book. The full extent of the graphing capabilities of R may be seen by searching the Internet for "R Graph Gallery" or "R Graph Catalog." Some modifications to the finer details of the base graphics system are introduced here, but sources such as Fox and Weisberg (2011), Mittal (2011), Murrell (2011), and Lillis (2014) should be consulted for a more thorough treatment of base graphics that will allow for wider customization of plots.

Required Packages for This Chapter

Functions used in this chapter require loading the packages shown below.[1]

```
> library(FSA)
> library(FSAdata)
> library(magrittr)
> library(dplyr)
> library(plotrix)
```

Data.Frames Used in this Chapter

Three data.frames from **FSAdata** are loaded below (see Section 2.1.4 for a description of `data()`) for use throughout this chapter. `BullTroutRML1` contains the fork lengths (`fl`) and weights (`mass`) of Bull Trout caught during two `eras` in Alberta, Canada (from Parker et al. 2007).

```
> data(BullTroutRML1)
> head(BullTroutRML1,n=3)
    fl mass     era
1   90   11 1977-79
2  180  107 1977-79
3  201  119 1977-79
```

`BullTroutRML2` contains the fork lengths and ages of Bull Trout from the same two eras but also recorded by lake of capture. These data are restricted below to fish from Harrison Lake for use in this chapter.

```
> data(BullTroutRML2)
> BullTroutRML2 %<>% filterD(lake=="Harrison")
> head(BullTroutRML2,n=3)
  age  fl    lake     era
1  14 459 Harrison 1977-80
2  12 449 Harrison 1977-80
3  10 471 Harrison 1977-80
```

BloaterLH contains the estimated number (in millions) of eggs deposited and relative abundance of age-3 Bloater by year in the Michigan waters of Lake Huron (from Schaeffer 2004).

```
> data(BloaterLH)
> head(BloaterLH,n=3)
  year   eggs    age3
1 1981 0.0402   5.143
2 1982 0.0602 154.286
3 1983 0.1205  65.143
```

3.1 Scatterplots

3.1.1 Simple Scatterplot

Scatterplots show the relationship between two quantitative variables. Scatterplots are created with plot(), with the variables to be plotted given in the first argument as a formula of the form y~x, where y and x generically represent the variables to be plotted on the y- and x-axes, respectively. The data.frame that contains these variables is given in data=.

Several optional arguments to plot() may be used to modify the default scatterplot. Labels for the x- and y-axes are given within quotes to xlab= and ylab=, respectively. The limits or range for the x- and y-axes are set with a vector that contains the minimum and maximum values for the axis in xlim= and ylim=, respectively. The default plotting symbol of an open circle is changed by setting pch= to a single number (e.g., 0="open square", 1="open circle").[2] A filled but outlined circle (i.e., pch=19) is used extensively in this book. The color of the plotted symbol is set with col=. Colors may be selected by name (e.g., "red" or "blue") or through color generating functions (e.g., rgb()).[3]

A scatterplot (Figure 3.1-Left) of weight versus fork length for Bull Trout, with modifications of the axis titles, axis limits, and plotting symbol, is constructed as shown below.

```
> plot(mass~fl,data=BullTroutRML1,ylim=c(0,1600),xlim=c(0,500),
      ylab="Weight (g)",xlab="Fork Length (mm)",pch=19)
```

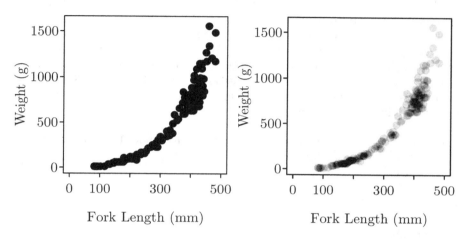

FIGURE 3.1. Scatterplot of Bull Trout weight versus fork length. The plot on the right illustrates the use of a semitransparent color to visualize the overplotting of points.

Large data sets that have relatively discrete variables (e.g., lengths) and little variability will often have considerable overplotting of points (e.g., Figure 3.1-Left). Overplotting reduces the information apparent in a plot because the density of individuals at a given point is not evident. The problem of overplotting can be mitigated by using a semitransparent color such that the apparent color becomes darker as more points are overplotted.

Semitransparent colors are constructed with rgb(), where the first three arguments are values between 0 and 1 that identify the intensity of the red, green, and blue portions of the custom color. The fourth argument to rgb() is a transparency value between 0 and 1, where the inverse is the number of overplotted points before the full color is evident. Figure 3.1-Right is an example where a solid black point (the value for the red, green, and blue portions are all zero) appears when four or more points are overplotted.

```
> plot(mass~fl,data=BullTroutRML1,ylim=c(0,1600),xlim=c(0,500),
      ylab="Weight (g)",xlab="Fork Length (mm)",
      pch=19,col=rgb(0,0,0,1/4))
```

3.1.2   Scatterplot with Different Symbols by Group

A fisheries scientist may want to construct a scatterplot with different symbols for different groups in the data (e.g., by sex, location, or time period). For example, one may want to show a scatterplot of Bull Trout weight versus length separated by the two eras of sampling.

The key to constructing such a plot is the creation of a vector of plotting characters or colors to use for *each individual* based on the particular group to which that individual belongs. These vectors can be manually created, but that is tedious and fraught with error. Fortunately, these vectors are easily created with the numeric codes that underlie each level of a factor variable (see Section 1.7.2) and recycling (see Section 1.7.1). This process is described below, first for a subset of six individuals to illustrate the process and then for all individuals in the data.frame.

```
> # select 6 individuals to illustrate the underlying process
> ( tmp <- BullTroutRML1[c(1:3,31:33),] )
     fl mass        era
1    90    11 1977-79
2   180   107 1977-79
3   201   119 1977-79
31  444   845    2001
32  439   798    2001
33  434   798    2001
```

A vector of symbol codes, colors, or both to be used for the multiple groups should be created first. These vectors should be at least as long as the number of groups to be plotted. For example, the vectors below identify the "plus sign" (pch=3) and "x" (pch=4) plotting characters and "black" and a shade of "gray" colors. With these declarations, individuals from the first era will be plotted with a black "plus sign" and individuals from the second era will be plotted with a gray "x."

```
> pchs <- c(3,4)
> cols <- c("black","gray60")
```

The user always sees the names of the levels for a factor variable.

```
> tmp$era
[1] 1977-79 1977-79 1977-79 2001    2001    2001
Levels: 1977-79 2001
```

However, as noted in Section 1.7.2, the levels are coded with numbers "behind-the-scenes." These codes are seen, if desired, by submitting the factor vector to as.numeric().

```
> as.numeric(tmp$era)
[1] 1 1 1 2 2 2
```

It is apparent, when the last two outputs are compared, that the first era ("1977-79") is coded as "1" and the second era ("2001") is coded as "2."

The underlying numeric codes for the factor levels are automatically used if a function or operator expects a numeric, but is given a factor vector. This is particularly useful when extracting values from another vector by position with []. For example, the following three lines of code extract the exact same information from the cols vector. The last line is much simpler and exploits the codes underlying the levels of the factor variable.

```
> cols[c(1,1,1,2,2,2)]
[1] "black"  "black"  "black"   "gray60" "gray60" "gray60"
> cols[as.numeric(tmp$era)]
[1] "black"  "black"  "black"   "gray60" "gray60" "gray60"
> cols[tmp$era]
[1] "black"  "black"  "black"   "gray60" "gray60" "gray60"
```

As this example shows, recycling makes it possible to extract the same position from a vector multiple times. The resulting vector, in this example, contains the color from the position in cols that corresponds to the underlying numeric codes in the factor vector. This vector of colors may then be given to col= in plot() to plot points with colors that are specific to the era recorded for each individual.

These principles were used below to select plotting characters from pchs and colors from cols that are specific to the two eras present in the Bull Trout data.frame (Figure 3.2). If col= had not been used, then the different symbols would both be in black (the default color). If pch= had not been used, then an open circle (the default plotting character) would be used but with the two colors. Finally, note that era alone, rather than BullTroutRML1$era, is used within [] because data= is used in plot().

```
> plot(mass~fl,data=BullTroutRML1,ylim=c(0,1600),xlim=c(0,500),
       ylab="Weight (g)",xlab="Fork Length (mm)",
       pch=pchs[era],col=cols[era])
```

A legend is added to an active plot with legend(). The first argument is the position where the legend should be placed. The position can be articulated in many ways (see ?legend), but the simplest is to give a named position such as "topleft", "topright", or "right". If a named position is used, then the legend is set away from the margins with inset= set equal to a proportion of the plotting area (e.g., 0.05 would move the legend 5% away from the named position). The group names are provided in legend= and are often extracted from the factor variable using levels(). The specific plotting symbols and

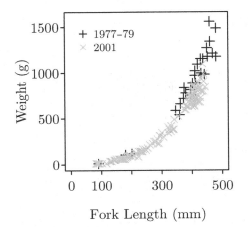

FIGURE 3.2. Scatterplot of Bull Trout weight versus fork length with different symbols and colors for the two different eras.

colors for the groups are given in `pch=` and `col=`, respectively. By default, the legend is encased in a box, which is removed with `bty="n"`. Finally, the relative size of the legend is changed with `cex=`, where values less than or greater than 1 perform a proportional reduction or enlarging, respectively, from the default size. The code below added the legend seen in Figure 3.2.

```
> legend("topleft",inset=0.05,legend=levels(BullTroutRML1$era),
      pch=pchs,col=cols,bty="n",cex=0.75)
```

3.2   Line Plots

3.2.1   Basic Line Plots

A line plot is also constructed with `plot()` by including `type="l"` for "line." The default solid line type may be modified by including a name or integer code in `lty=` (e.g., `1="solid"`, `2="dashed"`, and `3="dotted"`; see `?par` for a list of all line types). The line width is increased by including a value greater than 1 in `lwd=`. A line plot of the number of Bloater eggs in each sample year (Figure 3.3-Left) is constructed below.

```
> plot(eggs~year,data=BloaterLH,type="l",lwd=2,xlab="Year",
      xlim=c(1980,1996),ylab="Number of Eggs (Millions)")
```

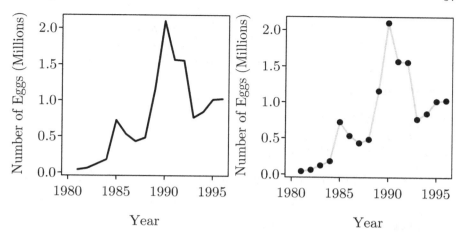

FIGURE 3.3. Line plot (Left) and line plot with points (Right) for the number of Bloater eggs in each sample year.

Both lines and points are plotted by first making the line plot as described above and then using `points()` to add points to the active plot. The `points()` function uses the same arguments as used in `plot()` to make the scatterplot.[4] The code below modified the line plot in Figure 3.3-Left by lightening the color of the line (i.e., `col="gray60"`) and including points (`cex=` was used to reduce the size of the points). The result is in Figure 3.3-Right.

```
> plot(eggs~year,data=BloaterLH,type="l",lwd=2,col="gray70",
       ylab="Number of Eggs (Millions)",xlab="Year",
       xlim=c(1980,1996))
> points(eggs~year,data=BloaterLH,pch=19,cex=0.75)
```

3.2.2   Superimposing a Line Plot on a Scatterplot

A line plot is superimposed onto a scatterplot by first creating the scatterplot as described above and then adding the line to the active plot with `lines()`. The `lines()` function uses the same arguments as used in `plot()` to make the line plot.[5]

For example, one might want to superimpose the mean lengths at each age on to a plot of length and age for each fish (Figure 3.4). To construct this plot, the mean length at each age (and, for use in later sections, the valid sample size and standard error using `validn()` and `se()`, respectively, from **FSA**) for Bull Trout must be computed as below.[6]

```
> sumBT <- BullTroutRML2 %>% group_by(age) %>%
    summarize(n=validn(fl),mnlen=mean(fl,na.rm=TRUE),
              selen=se(fl,na.rm=TRUE)) %>%
    as.data.frame()
> head(sumBT,n=3)
  age n     mnlen      selen
1   0 3  37.33333   9.134793
2   1 4  96.25000  12.092525
3   2 5 178.60000  10.181356
```

The scatterplot of length versus age is then constructed and the mean lengths at each age superimposed as a solid black line.

```
> plot(fl~age,data=BullTroutRML2,pch=19,col=rgb(0,0,0,1/3),
       ylab="Fork Length (mm)",xlab="Age (yrs)")
> lines(mnlen~age,data=sumBT,lwd=2)
```

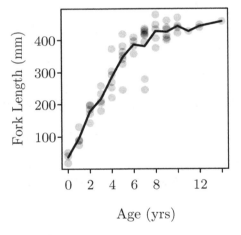

FIGURE 3.4. A scatterplot of Bull Trout lengths versus age with the mean lengths-at-age shown with the solid line.

Superimposing best-fit lines or curves is illustrated in Section 3.5.1.

3.3   Histograms

Histograms show the distribution of a quantitative variable and may be constructed with hist() from base R. However, this function uses "left-exclusive, right-inclusive" bins (e.g., a value of 70 is placed in the 60-70 bin rather than the 70-80 bin) which are not usually used by fisheries scientists, includes a default main title which is typically removed in fisheries applications, defaults to an x-axis that appears to "float" below the bars, and is not conducive to constructing histograms separated by levels of a factor variable. A modified version of hist() from **FSA** efficiently rectifies these four issues.[7] Variables **must be** given in a formula (note use of ~) to hist() from **FSA**, whereas hist() in base R does not work with formulae. The hist() from **FSA** is used exclusively in this book.

3.3.1 Single Histograms

A histogram for a single group requires a formula of the form ~quant as the first argument, where quant is a quantitative variable, and the data.frame that contains quant in data=. Characteristics of the histogram may be modified as described for plot() (e.g., xlab=, ylab=, xlim=, ylim=, col=). A histogram of Bull Trout weight (Figure 3.5-Left) is constructed below.

```
> hist(~mass,data=BullTroutRML1,xlab="Weight (g)",
        ylim=c(0,50),xlim=c(0,1600))
```

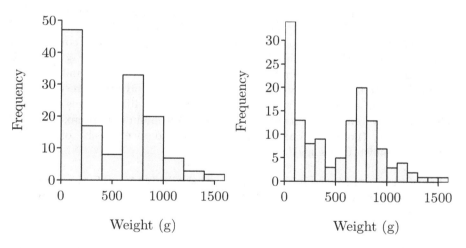

FIGURE 3.5. Histograms of the weight of Bull Trout. The histogram on the right uses custom bin widths.

The default choice for bins likely does not meet the needs of the fisheries scientist and will need to be modified. The starting (i.e., minimum) values of the bins can be entered into a vector and then assigned to breaks= in hist(). If equal bin widths are used, then a sequence of values between a minimum (first argument) and maximum (second argument) value with a certain "step" (third argument) is constructed with `seq()`. In the context of a histogram, the user must make sure that the minimum and maximum values in this sequence are smaller and greater, respectively, than the minimum and maximum observed values in the data to be plotted. The code below first finds the range of observed weights with `range()` (using `na.rm=TRUE` to remove any missing weights) and then uses the custom bins shown in Figure 3.5-Right.

```
> range(BullTroutRML1$mass,na.rm=TRUE)
[1]    11 1559
> hist(~mass,data=BullTroutRML1,xlab="Weight (g)",
       xlim=c(0,1600),breaks=seq(0,1600,100))
```

3.3.2   Histograms for Multiple Groups

Histograms separated by the levels of a factor variable (i.e., by group) are constructed with a formula of the form `quant~factor` in `hist()`, where `factor` is the factor variable that identifies group membership. By default, the separate histograms will use the same bins (or breaks) and the same limits for the y-axis. These default options allow ease of comparison among the groups, but those options may be "turned off" with `same.breaks=FALSE` and `same.ylim=FALSE`, respectively. The bins (or breaks) are controlled with `breaks=` as before, but the limits for the y-axis are controlled with `ymax=` when `same.ylim=TRUE`. Each histogram will have a main title that is constructed from the levels of the factor variable. A prefix is appended to these titles by including that prefix in `pre.main=`.[8] Finally, the number of rows and columns for the multiple histograms are controlled with `nrow=` and `ncol=`, respectively.

A histogram of the weight of Bull Trout from the different eras sampled, using the same bin breaks, same y-axis limits, and main title prefixes is constructed below (note that `cex.main=` was used to reduce the size of the plot titles) and shown in Figure 3.6.

```
> hist(mass~era,data=BullTroutRML1,xlab="Weight (g)",ymax=35,
       breaks=seq(0,1600,100),pre.main="Era = ",cex.main=0.9)
```

3.4   Bar Plots

3.4.1   Frequency of Individuals in Groups

Bar plots show the frequency of individuals in each level of a factor variable. Bar plots are constructed with `barplot()`, which requires a table of frequencies, rather than a variable, as the first argument. A table of frequencies is constructed with `xtabs()`, which takes a formula of the form `~factor` and

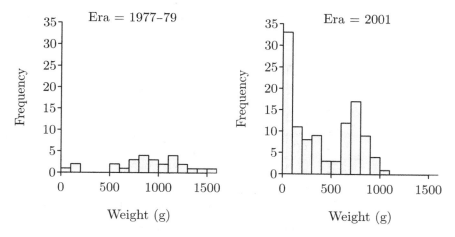

FIGURE 3.6. Histograms of the weight of Bull Trout by era.

a data.frame that contains the `factor` variable in `data=`. For example, the number of sampled Bull Trout in each era is summarized below.

```
> ( eraBT <- xtabs(~era,BullTroutRML1) )
era
1977-79    2001
     27     110
```

The result of `xtabs()` must be saved to an object for submission as the first argument to `barplot()`. Several of the optional arguments for `plot()` are also used with `barplot()`. The result of the code below is shown in Figure 3.7.

```
> barplot(eraBT,ylab="Number of Fish",xlab="Era",ylim=c(0,120))
```

3.4.2 Other Values

Fisheries scientists also commonly use a bar plot to plot values other than frequencies against levels. Some of these plots may be constructed with `plotH()` from **plotrix**, which takes a formula of the form `quant~x`, where x is either a discrete quantitative or factor variable to be plotted on the x-axis. The same suite of optional arguments from `plot()` may also be used here. For example, the mean lengths at each age in `sumBT` (from before) are plotted in Figure 3.8.

```
> plotH(mnlen~age,data=sumBT,ylim=c(0,600),
        ylab="Mean Fork Length (mm)",xlab="Age (years)")
```

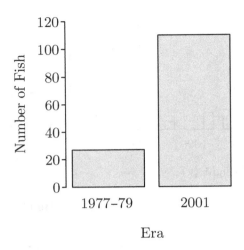

FIGURE 3.7. Bar plot of number of Bull Trout captured by era.

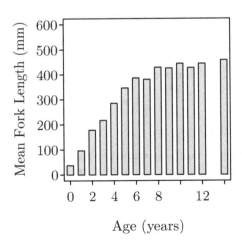

FIGURE 3.8. Bar plot of Bull Trout mean length-at-age.

3.5 Fitted Model Plots

3.5.1 Line or Curve

In many instances a fisheries scientist will want a graphic that illustrates the fit of a model to data. While there are many possible ways to superimpose model fits, a method that is general and extensible, though slightly more work in the simplest examples, is demonstrated in this section. This method, used in several ensuing chapters, is illustrated here for visualizing the fit of a linear regression to log-log transformed weight-length data for Bull Trout. Linear regression of weight-length data is discussed in more detail in Chapter 7. However, for the purpose here, note that lm() takes a formula of the form y~x as the first argument, where y generically represents the response (or dependent) variable and x generically represents the explanatory (or independent) variable, and the corresponding data.frame in data=. The natural log of both variables are added to the data.frame first.

```
> BullTroutRML1 %<>% mutate(logw=log(mass),logl=log(fl))
> lm1 <- lm(logw~logl,data=BullTroutRML1)            # fit SLR
```

The fit of any continuous model is visualized by first creating a vector of values that cover the range of observed values for the explanatory variable. The vector of explanatory values is easily created with seq(). In this context, seq() takes the minimum and maximum values as the first two arguments and the number of values in the sequence in length.out=. Large values in length.out= will produce a smoother curve but will take longer to compute. A vector of 99 log lengths for the Bull Trout model is created below. In this example, note that range() is used to find the range of observed log length values with the result assigned to an object. In seq(), the minimum and maximum values are then extracted from the first and second positions of the range() object.[9]

```
> ( rng <- range(BullTroutRML1$logl) )
[1] 4.406719 6.173786
> xs <- seq(rng[1],rng[2],length.out=99)
```

The fitted model is then used to predict values of the response variable for each of these values of the explanatory variable. The predicted values are obtained by creating an expression of the model using the parameter estimates from the model fit and the sequence of values for the explanatory variable just created. For a linear model (i.e., the lm() object created above), the parameter estimates are extracted with coef(). It is most efficient to save the parameter estimates into their own object.

```
> ( ps <- coef(lm1) )
(Intercept)        logl
 -10.317873    2.822465
```

Predicted log weights for the Bull Trout are then found using an expression for the linear model that includes the slope and intercept from the `coef()` object and the vector of log length values.[10]

```
> ys <- ps[["(Intercept)"]]+ps[["logl"]]*xs
```

The fitted-line plot (Figure 3.9-Left) is then constructed by first making a scatterplot of the original data with `plot()`.

```
> plot(logw~logl,data=BullTroutRML1,pch=19,col=rgb(0,0,0,1/4),
      ylab="Log Weight (g)",xlab="Log Length (mm)")
```

The sequence of explanatory values and the corresponding predicted values are then given to `lines()` to superimpose the model. The `lines()` function takes a formula of the form `quant~x` and, if the data are in a data.frame (which is not the case here), then that data.frame would be given to `data=`.

```
> lines(ys~xs,lwd=2)
```

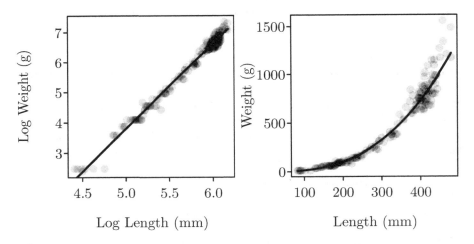

FIGURE 3.9. Scatterplot of Bull Trout log weight versus log length (Left) and weight versus length (Right) with the best-fit line superimposed.

This method for superimposing the fitted line from the model is general enough that the same rationale may be used to superimpose the back-transformed fitted model onto the original data (Figure 3.9-Right). In the example below, it should be noted that the back-transformed model is $y = ax^b$

where b is the slope and a is the back-transformed intercept (i.e., $e^{intercept}$ where exp() is used to raise its argument to the power of e) from the fit of the transformed model. This equation is used in the construction of the predicted values (i.e., the ys).

```
> # range of observed fork lengths
> rng <- range(BullTroutRML1$fl)
> # vector of 99 fork lengths
> xs <- seq(rng[1],rng[2],length.out=99)
> # vector of 99 predicted weights at each fork length
> ys <- exp(ps[["(Intercept)"]])*xs^ps[["logl"]]
> # plot the points
> plot(mass~fl,data=BullTroutRML1,pch=19,col=rgb(0,0,0,1/4),
        ylab="Weight (g)",xlab="Length (mm)")
> # superimpose the curve
> lines(ys~xs,lwd=2)
```

An alternative method is to use curve(). The first argument to curve() is an expression of the right-hand side of the model as above, but with an "x" in place of the actual explanatory variable. The domain over which the equation should be plotted starts at the value in from= and ends with the value in to=. The curve is superimposed onto an active scatterplot if add=TRUE is used. A plot similar to Figure 3.9-Right would be constructed using curve() as shown below. Note that the same back-transformed model used above forms the first argument to curve() in this example.

```
> plot(mass~fl,data=BullTroutRML1,pch=19,col=rgb(0,0,0,1/4),
        ylab="Weight (g)",xlab="Length (mm)")
> curve(exp(ps[["(Intercept)"]])*x^ps[["logl"]],add=TRUE,
        from=rng[1],to=rng[2],lwd=2)
```

The method using curve() is slightly simpler when only the best-fit line is superimposed, but it is less general and cannot be used for some of the more advanced applications in later chapters.

3.5.2 Means Plot with Vertical Intervals

A plot of group means with confidence intervals (Figure 3.10) is a useful summary following a statistical test of means (e.g., a one-way ANOVA; see Section 8.3.2). This type of plot is easily constructed with plotCI() from **plotrix**, if a data.frame with the group means and associated confidence intervals exists.

A data.frame with the mean, standard error, and valid sample size of Bull Trout at each age (i.e., sumBT) was constructed in Section 3.2.2. Approximate confidence intervals are constructed from these data by adding and subtracting the standard error multiplied by a critical value from the t-distribution from the mean. These critical values are obtained from `qt()` with the probability level[11] as the first argument and the corresponding degrees-of-freedom in `df=`. Note that a warning may be given, as in this example, when a sample size equals 1 because the degrees-of-freedom will then be zero.

```
> conf.level <- 0.95
> tcrit <- qt(0.5+conf.level/2,df=sumBT$n-1)

Warning in qt(0.5 + conf.level/2, df = sumBT$n - 1): NaNs
produced
```

The lower and upper bounds for the confidence interval are added to the data.frame with `mutate()` from **dplyr**.

```
> sumBT %<>% mutate(LCI=mnlen-tcrit*selen,UCI=mnlen+tcrit*selen)
> headtail(sumBT,n=2)
   age n     mnlen     selen        LCI        UCI
1    0 3  37.33333  9.134793  -1.970509   76.63718
2    1 4  96.25000 12.092525  57.766188  134.73381
13  12 2 444.50000  4.500000 387.322079  501.67792
14  14 1 459.00000        NA         NA         NA
```

The x- and y-axis values for plotting the means are given in the first two arguments and the upper and lower endpoints of the vertical intervals are given in `ui=` and `li=`, respectively, of `plotCI()`. For example, the code below uses the results in `sumBT` to plot the mean length at each age with confidence intervals. The `cex=0.7` was used to make the points smaller so that the intervals were more visible.

```
> plotCI(sumBT$age,sumBT$mnlen,
         li=sumBT$LCI,ui=sumBT$UCI,pch=19,cex=0.7,
         xlab="Age (yrs)",ylab="Fork Length (mm)")
```

Other examples of using `plotCI()` to add intervals to a plot are in Section 6.3.7 and Section 8.3.2.

3.6   Some Finer Control of Plots

3.6.1   Plotting Parameter Options

The specific details of base graphics are controlled through a large number of plotting parameter options. The list of current settings for the plotting parameter options is seen with `par()` (i.e., with no arguments).

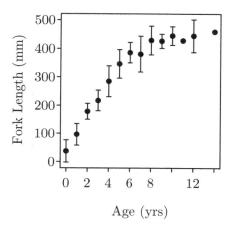

FIGURE 3.10. Mean length-at-age, with corresponding 95% confidence intervals, for Bull Trout.

```
> par()[1:4]   # only 4 to save space (remove [1:4] to see all)
$xlog
[1] FALSE

$ylog
[1] FALSE

$adj
[1] 0.5

$ann
[1] TRUE
```

The meanings of each option are described in detail in the help documentation for par(). These options are changed by including the option as an argument to par() prior to the construction of a plot. Some of the plotting parameter options, however, may be set within functions like plot(), hist(), points(), and lines(). Examples of how to change the most common options used by the fisheries scientists (and used in this book) are described below.

The size of figure margins are measured in "lines" and can be set separately for the bottom, left, top, and right sides (in that order, respectively) with mar=.[12] The default values for the margins are c(5.1,4.1,4.1,2.1) (Figure 3.11). The axes titles, values, and lines, respectively, are plotted on the figure margin lines given in mgp=, which default to c(3,1,0) (Figure 3.11) (note how the axis titles align with "line=3" and the value labels align with "line=1").

Tick mark lengths are set with tcl=, where a negative number produces ticks outside the plotting region and positive numbers produce ticks inside the

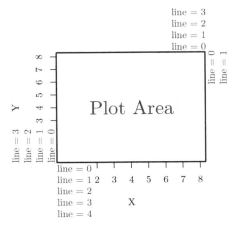

FIGURE 3.11. Plot schematic illustrating the lines in each margin of the plotting area. Note that `mar=c(5.1,4.1,4.1,2.1)` and `mgp=c(3,1,0)`.

plotting region (default is `tcl=-0.5`). By default, the value labels are parallel to the axes. However, the value labels will always be horizontal to the plot when `las=1` is used. The size of the text and plotting symbols is adjusted with `cex=` where values different than 1 represent a proportional decrease (less than 1)·or increase (greater than 1) relative to the default size. Alternatively, the axis titles and value labels may be modified separately with `cex.axis=` and `cex.lab=`, respectively.

Two other plotting parameter options that are useful to fisheries scientists are the ability to change how the axes relate to each other and change if points outside of the plotting region are clipped. By default, R finds "pretty labels" for the axes by extending the axes 4% beyond the observed data at both ends. Thus, the axes may not cross at the minimum values in `xlim=` and `ylim=`. Including `xaxs="i"` and `yaxs="i"` in `par()` will force the axes to cross at the minimum values for each axis. With these settings, however, the points that fall on either axis will appear "clipped." The full point is seen by setting `xpd=TRUE`. Use of these options is illustrated below with the result shown in Figure 3.12.

```
> par(xaxs="i",yaxs="i",xpd=TRUE)
> plot(mass~fl,data=BullTroutRML1,pch=19,col=rgb(0,0,0,1/4),
        ylab="Weight (g)",xlab="Length (mm)")
```

Most plots in this book have narrower margins, axis titles and value labels that are closer to the axes, smaller ticks, axis value labels that are always horizontal, and axis titles and value labels that are slightly smaller than the default. The plotting parameter options used in this book are shown in the Preface.

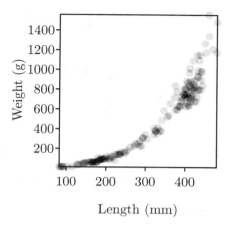

FIGURE 3.12. Scatterplot of Bull Trout weight versus fork length illustrating modified axis settings.

While not used in this book, the *font face* is changed with font=, whereas the *font family* is changed with family= in par(). The default font face is plain text (i.e., font=1) but font=2 will use bold, font=3 will use italics, and font=4 will use bold and italics for all text in the plot. Use font.lab= and font.axis= to change only the axis titles and the axis value labels, respectively. Available font families depend on the type of graphics device and operating system being used. However, typical families include "serif", "sans", and "mono". The scatterplot (Figure 3.13-Left) constructed below illustrates the use of different fonts for the axis and axis value labels.

```
> par(font.lab=2,font.axis=4,family="sans")
> plot(mass~fl,data=BullTroutRML1,
        ylab="Mass (g)",xlab="Fork Length (mm)",
        ylim=c(0,1600),xlim=c(0,500),pch=19,col=rgb(0,0,0,1/4))
```

Axis value labels may be omitted from a plot if the algorithm determines that there is not enough room for those labels (e.g., 200 and 400 are missing from the x-axis in Figure 3.13-Left). Labels may be added to an axis of an active plot with axis(). The first argument is the side of the plot on which to add the axis, beginning with the x-axis at 1 and continuing clockwise to the right-side y-axis at 4. The second argument is a vector of values at which to put the labels. Labels are given in the third argument, but if no third argument is given, then the labels will be the values given in the second argument. The code below added the missing x-axis labels to the previous scatterplot (Figure 3.13-Right).

```
> axis(1,c(200,400))
```

Introductory Fisheries Analyses with R

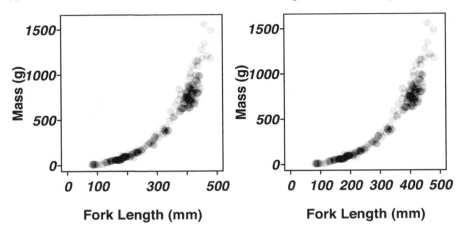

FIGURE 3.13. Scatterplot of Bull Trout weight versus length illustrating different types of fonts. The plot on the right has a more complete x-axis.

3.6.2 Adding Text to a Plot

Text may be added at specific coordinates of the active plot with `text()`. In its simplest form, `text()` takes vectors for the x- and y-coordinates at which to plot the text in the first two arguments and a vector of "text" to plot at those coordinates as the third argument. For example, the code below constructs a scatterplot and then places two simple notes on the plot (Figure 3.14).

```
> plot(selen~mnlen,data=sumBT,pch=19,col=rgb(0,0,0,1/2),
        ylab="SE Fork Length (mm)",xlab="Mean Fork Length (mm)",
        xlim=c(0,500),ylim=c(0,40))
> text(c(225,300),c(35,3),c("Parabolic Shape?","Why Low?"))
```

The coordinates can also be variables from a data.frame. For example, the code below creates the same scatterplot as above, but with the color of the points set to white to make them "invisible." The corresponding age value is then plotted at each point using `text()` (Figure 3.15-Left).

```
> plot(selen~mnlen,data=sumBT,pch=19,col="white",
        ylab="SE Fork Length (mm)",xlab="Mean Fork Length (mm)",
        xlim=c(0,500),ylim=c(0,40))
> text(sumBT$mnlen,sumBT$selen,sumBT$age,cex=0.8)
```

Text is placed to the side of the x- and y-coordinate positions by including a value in `pos=`, where 1 is below, 2 is left, 3 is above, and 4 is right of the position. For example all text values are placed above the points with the

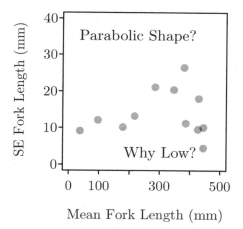

FIGURE 3.14. Scatterplot of the standard error versus mean length-at-age of Bull Trout.

code below (Figure 3.15-Right). Individual positions can be used by creating a vector of position numbers that is as long as the number of points plotted.

```
> plot(selen~mnlen,data=sumBT,pch=19,col=rgb(0,0,0,1/2),
        ylab="SE Fork Length (mm)",xlab="Mean Fork Length (mm)",
        xlim=c(0,500),ylim=c(0,40))
> text(sumBT$mnlen,sumBT$selen,sumBT$age,pos=3,cex=0.8)
```

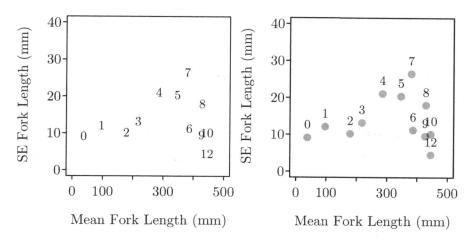

FIGURE 3.15. Scatterplot of the standard error versus mean length-at-age of Bull Trout, with age labels as the points (Left) and plotted above the points (Right).

Of course, combinations of the methods illustrated here can be used. For example, one could give coordinates with the variables from a data.frame but manually create a vector of labels. This flexibility is used in subsequent chapters (e.g., Section 8.3.2).

3.6.3   Simple Placement of Multiple Plots

At times it may be useful to place two plots side-by-side or one-over-the-other, or to place four plots in a two-by-two array. These simple arrays of plots are accomplished by using mfrow= or mfcol= within par() prior to making the plots. Both of these arguments accept a vector of two values that indicate the number of rows and number of columns in an array for positioning the plots. The only difference between mfrow= or mfcol= is the order in which the plots are placed into the array — by row and then columns for mfrow= or by columns and then rows for mfcol=. For example, the side-by-side (i.e., one row, two columns) plot in Figure 3.16 is constructed with the following code.

```
> par(mfrow=c(1,2))
> plot(mass~fl,data=BullTroutRML1,
        ylab="Mass (g)",xlab="Fork Length (mm)",
        ylim=c(0,1600),xlim=c(80,500),pch=19,col=rgb(0,0,0,1/4))
> hist(~fl,data=BullTroutRML1,xlab="Fork Length (mm)",
        breaks=seq(80,500,10))
```

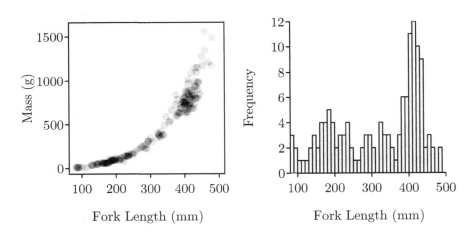

FIGURE 3.16. Scatterplot of mass versus fork length (Left) and histogram of fork length (Right) for Bull Trout.

A one-over-the-other plot (i.e., two rows and one column) would use par(mfrow=c(2,1)) and the two-by-two array of plots entered in the first column and then the second column would use `par(mfcol=c(2,2))`. More complex layouts (e.g., plot in first row spans two columns, unequal sized plots, or common x- and y-axis labels) are described in the online supplement.

3.7   Saving or Exporting Plots

Plots made in R may be saved to an external file in a variety of formats by calling the proper device, making the plot, and then closing the device with `dev.off()`. Two common formats are *portable document format* (PDF) and *portable network graphics* (PNG) files. Generally, PDF is best for printing because it is vector-based and scale-independent, which means that it can be resized without creating pixelated or jagged lines and text. Plots saved in PDF are high quality, can be viewed with commonly available viewers across all operating systems, and can be integrated with other PDF documents. Plots in PNG format are best used for display within web pages (PDF does not integrate well with HTML) and importing into a word processor.[13]

A PDF file is created with `pdf()`. The first argument is the name for the external file in quotes (should end with ".pdf"). The width and height in inches of the resulting plot are set with `width=` and `height=`, respectively, with both values set to 7 by default. Several other options, including the size of the text, font, and background color, can be set in `pdf()` (see the `pdf()` help documentation for more details on options). For example, the following code will produce a histogram that is 5 inches wide by 4 inches tall in an external PDF file called "ExHist.pdf" in the current working directory (see Section 2.1.1 for how to view and change the current working directory).

```
> pdf("ExHist.pdf",width=5,height=4)
> hist(~mass,data=BullTroutRML1,xlab="mass (g)")
> dev.off()            # required to produce the PDF file
```

A PNG file is created similarly with `png()`. The first argument is again the name for the external file (should end with ".png"). By default, the size of the plot is set in pixels. However, if `units="in"`, then the width and height of the resulting plot is set in inches with `width=` and `height=`, respectively. Additionally, the default resolution is 72 pixels per inch, but this can be increased by including a larger integer in `res=`. For example, the following code creates a 5 inches wide by 4 inches tall histogram, with an increased resolution, in the "ExHist.png" external PNG file.

```
> png("ExHist.png",width=5,height=4,units="in",res=144)
> hist(~mass,data=BullTroutRML1,xlab="mass (g)")
> dev.off()            # required to produce the PNG file
```

A variety of other formats for outputting graphics exist (e.g., `jpeg()`, `bmp()`, `tiff()`, `svg()`, `win.metafile()`) that have similar arguments to `pdf()` and `png()` (specifics can be found in the help documentation for each function). Additionally, the examples above provide only a simple introduction for the most common uses by fisheries scientists; please consult the help documentation and numerous examples available via the Internet for discussions of more advanced functionality.

Finally, many users will copy plots directly from R or RStudio and paste them into a word processor or other software. This tends to produce very low quality graphics. Additionally, using `pdf()` and `png()` within a script (see Section 1.10) means that the exact plot can be reproduced in the future. Thus, I suggest using one of these functions to save your plots in external files.

Notes

[1] Data manipulations in this chapter require functions from **magrittr** and **dplyr**, which are fully described in Chapter 2.

[2] The `pch=` argument is for "plotting character." The plotting symbols that correspond with the integers used in `pch=` are described in the help documentation for `points()`. In addition, `show.pch()` from **Hmisc** graphically shows the available plotting symbols.

[3] Names of available colors may be seen with `colors()`. Other color generating functions are `hsv()`, `hcl()`, `gray()`, and `rainbow()`.

[4] The `points()` function does not use `xlab=`, `ylab=`, `xlim=`, or `ylim=` as these were set when the active plot was created.

[5] The `lines()` function does not use `xlab=`, `ylab=`, `xlim=`, or `ylim=` as these were set when the active plot was created.

[6] See Section 2.5 for descriptions of `group_by()`, `summarize()`, and `as.data.frame()`.

[7] The `hist()` function in base R can use "left-inclusive, right-exclusive" bins by including `right=FALSE` and the main title may be removed with `main=""`. These are the default options with `hist()` provided by **FSA**.

[8] If `pre.main=NULL`, then no main title will be printed above each histogram.

[9] Some users may find it easier to extract the minimum and maximum values from the `range()` vector with `min()` and `max()`. In this case, the second line would look like `xs <- seq(min(rng),max(rng),length.out=99)`.

[10] The predicted values from a simple linear regression model are more easily constructed with `predict()`. Specifically, in this case, use `ys <- predict(lm1,data.frame(logl=xs))`. The method shown here is more general to a wider variety of applications.

[11] The probability level for a $100C\%$ confidence interval is $0.5 + \frac{C}{2}$ because the area in both tails of the distribution must be $\frac{1-C}{2}$.

[12] The size of the figure margins can also be set with `mai=` using the same arguments as `mar=`, but using values that are in inches. The size of a "line" in inches is found with `par("mai")/par("mar")`. It is generally best to use `mar=` because several of the other options (e.g., `mgp=`) and functions (e.g., `mtext()`) use "lines" as their measure.

[13] The relatively new SVG format is a scaleable vector graphic, like PDFs, that can be used in web pages that are now supported by most modern web browsers.

4

Age Comparisons

A foundation for using R for fisheries analyses was constructed in Chapters 1–3. Methods in this and Chapter 5 will build upon this foundation to assess and prepare age data for analysis in later chapters.

Information about the age of fish is used to determine rates of mortality (Chapter 11), growth (Chapter 12), and maturation, which are key components of population models used to make fisheries management decisions. Ages of fish are often estimated by assessing patterns found in calcified structures such as scales, fin rays or spines, and otoliths. Ages estimated from calcified structures may not equal the true age of the fish, may not be the same among different technicians, or may not be the same between structures. Given the importance of age data in fisheries management, it is critical to understand sources of variability, precision, and bias in age estimates.

Four terms are commonly used when comparing estimated ages to true ages, among structures, among technicians, or among times by the same technician. *Accuracy* relates whether the estimated age equals the true age of the fish. If the estimated age equals the true age throughout the lifespan of a species, then that age structure is said to have been *validated*. If the estimated age tends to over or under estimate the true age, likely for a portion of the lifespan of the species, then the calcified structure is *biased*. However, *bias* also describes the situation where one set of estimated ages over or under estimates another set of estimated ages for the same individuals. Finally, *precision* measures the repeatability of age estimates on the same individuals and may occur with or without bias. Precision is usually measured among multiple technicians interpreting the same structure or among multiple interpretations of the same structure by the same technician. Preferably, ages are determined from a structure that shows no bias from the true age (i.e., validated) with a high level of repeatability among technicians or over time (i.e., precise).

Required Packages for This Chapter

Functions used in this chapter require to load **FSA**.

```
> library(FSA)
```

4.1 Data Requirements

The true age of each fish must be known to determine the accuracy of age estimates. Bias among structures or technicians is determined from paired age estimates for the same fish. Finally, precision among structures or technicians is examined with two *or more* age estimates for the same fish.

As an example, McBride et al. (2005) examined the validity of scales to estimate the age of American Shad. The true ages for fish in their sample were known because the Shad had been marked prior to being stocked. Additionally, 13 biologists twice (independently) estimated the age from scales for each fish. The known age of the fish (**trueAge**) and the age estimates from three of the 13 biologists are available in *ShadCR.csv*. The estimated age variables are labeled with **ager**, a letter for the three biologists (A, B, or C) and a number for which time the scale was interpreted (1 or 2). Some biologists chose not to assign an age to some scales and, thus, those data are missing (shown as NA values).

```
> shad <- read.csv("ShadCR.csv")
> headtail(shad)
    fishID trueAge agerA1 agerA2 agerB1 agerB2 agerC1 agerC2
1   00-LR2       5      5      3     NA     NA      4      3
2  00-LR20       4      3      3      5      4      2      2
3  01-LR30       6      5      5     NA     NA      6      6
51 03-3293       8      7      5     NA     NA      6      4
52  00-LR8       5      7      7     NA     NA      4      4
53 LR54-00       5      6      6     NA     NA      6      3
```

The shad data will be used to examine (1) accuracy by comparing the estimated ages from a biologist to the known ages (for simplicity, only the first estimate for a biologist will be used), (2) bias between biologists by comparing the age estimates for two biologists (again, only using the first estimates), and (3) precision by comparing between the two estimates by the same biologist and also by comparing among the first estimates of all three biologists.

4.2 Age Bias Plot

Campana et al. (1995) introduced the *age bias plot* to identify systematic bias between two sets of ages from the same fish. To construct an age bias plot, one of the two sets of ages must be considered as a "reference." If one is comparing estimated ages to true ages, then the true ages would be the reference set. However, if both ages are estimated, then the ages that are presumed to be most accurate (e.g., from the more experienced technician) should serve as the reference set.

In an age bias plot (Figure 4.1), the mean age and associated confidence interval computed from the nonreference ages for each reference age is plotted against the corresponding reference age. The plot is augmented with a 1:1 agreement line (i.e., points where the mean nonreference and reference ages are equal). Confidence intervals that do not capture the agreement line indicate a difference between the mean nonreference and reference ages.

Constructing an age bias plot is a two-step process. First, the mean and associated confidence intervals of the nonreference ages for each reference age are computed with `ageBias()` from **FSA**. The first argument is a formula of the form `nonrefvar~refvar`, with `nonrefvar` and `refvar` generically representing the variables that contain the nonreference and reference ages, respectively. The data.frame that contains these variables is given in `data=`. Labels for the x- (reference) and y- (nonreference) axes may be given in `ref.lab=` and `nref.lab=`, respectively. Second, the age bias plot is constructed by submitting the `ageBias()` object to `plot()`.

```
> ab.tA1 <- ageBias(agerA1~trueAge,data=shad,
                    ref.lab="True Age",nref.lab="Ager A")
> plot(ab.tA1,col.CIsig="black")
```

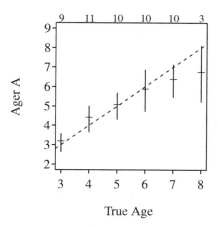

FIGURE 4.1. Age bias plot for comparing the mean estimated ages from Ager A to the true ages of American Shad.

Confidence intervals are only shown in the plot for reference ages with a sample size at least equal to the value in `min.n.ci=` (defaults to 3) in `ageBias()`. The confidence intervals are color-coded by default to highlight whether the mean of the nonreference ages is significantly different from the corresponding reference age. Significance is determined by comparing each p-value from a one-sample t-test adjusted for multiple comparisons to the `sig.level=` value in the original `ageBias()` call (=0.05 by default).[1] The confidence intervals in the age bias plot are *unadjusted* for multiple comparisons; thus, some confidence intervals may appear to not capture the agreement line but also not be coded as being significantly different. Color coding for the significance tests may be avoided by setting `col.CI=` (defaults to `"black"`) and `col.CIsig=` to the same color in `plot()`, as was done in Figure 4.1 for the purpose of printing.

The default age bias plot does not show the range of nonreference ages for each reference age. These ranges are included (Figure 4.2-Left) with `show.range=TRUE` in `plot()`.

```
> plot(ab.tA1,col.CIsig="black",show.range=TRUE)
```

An alternative to showing the range of values is to show the individual data points on the age bias plot with `show.pts=TRUE` in `plot()`. However, there is generally considerable overplotting of points due to the discrete nature of age data. One method to handle overplotting is to plot a datum as a semitransparent point (e.g., see Section 3.1). The number of overplotted points required to make the point appear completely dark is controlled with `transparency=` in `plot()`. For example, using `transparency=1/6` would require six overplotted points to create a completely dark point (Figure 4.2-Right).

```
> plot(ab.tA1,col.CIsig="black",show.pts=TRUE,transparency=1/6)
```

The statistics underlying an age bias plot may be viewed by submitting the `ageBias()` object to `summary()` with `what="bias"`.[2]

```
> summary(ab.tA1,what="bias")
```

trueAge	n	min	max	mean	SE	t	adj.p	sig	LCI	UCI
3	9	2	4	3.11	0.200	0.555	1.000	FALSE	2.65	3.57
4	11	3	6	4.33	0.302	1.106	1.000	FALSE	3.66	5.01
5	10	4	7	5.00	0.298	0.000	1.000	FALSE	4.33	5.67
6	10	3	8	5.80	0.467	-0.429	1.000	FALSE	4.74	6.86
7	10	5	8	6.30	0.367	-1.909	0.443	FALSE	5.47	7.13
8	3	6	7	6.67	0.333	-4.000	0.343	FALSE	5.23	8.10

Muir et al. (2008) modified the age bias plot by plotting the differences in paired ages on the y-axis. This modification may simplify interpretation when a wide range of ages are present in the sample. This modified age bias plot (Figure 4.3-Left) is constructed by including `difference=TRUE` in `plot()`.

```
> plot(ab.tA1,col.CIsig="black",difference=TRUE)
```

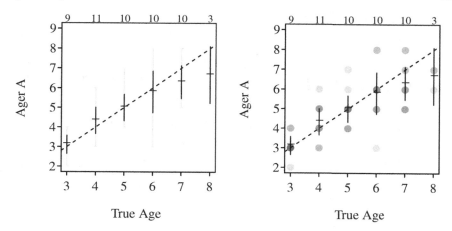

FIGURE 4.2. Age bias plots for comparing the mean estimated ages from Ager A to the true ages of American Shad. The gray intervals on the left plot represent the range of estimated ages for Ager A at each true age. The darkness of the points on the right plot is proportional to the number of individuals at that point.

While it is not an age bias plot as defined by Campana et al. (1995), it is common to plot the numbers of individuals that would be overplotted at each point (Figure 4.3-Right). This plot is constructed by including `what="number"` in `plot()`. It may be prudent to lighten the agreement line with `col.agree=` so that the numbers can be more easily read.

```
> plot(ab.tA1,what="number",col.agree="gray50")
```

4.3 Bias Metrics

Modern[3] statistical tests for bias focus on whether an age-agreement table is symmetric around the main diagonal. An age-agreement table is a contingency table that shows how two sets of ages are related. The main diagonal in an age-agreement table represents the individuals for which the two ages are the same (analogous to the 1:1 line in the age bias plot). If there is no systematic bias between the two ages, then individuals where the paired ages disagree are equally likely to be above or below this main diagonal. Thus, the age-agreement table is *symmetric* around the main diagonal if there is no systematic bias between the two paired ages.

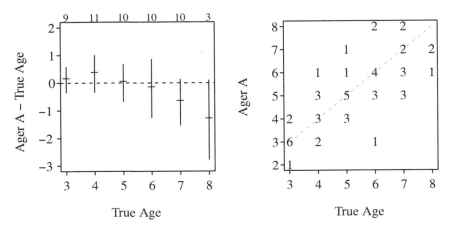

FIGURE 4.3. Alternative age bias plots for comparing the mean estimated ages from Ager A to the true ages of American Shad. The left plot shows the mean difference in the estimated and true ages on the y-axis. The right plot shows numbers that represent the number of fish at each point.

Evans and Hoenig (1998) described three tests for assessing the symmetry of an age-agreement table. The *McNemar test* (also known as *maximally pooled*; McNemar 1947; Hoenig et al. 1995) tests if the number of individuals below the main diagonal ($j - i < 0$, where i is the row number and j is the column number of the age-agreement table) is equal to the number of individuals above the main diagonal ($j - i > 0$; Figure 4.4). The McNemar test does not take into account where the values are relative to the main diagonal beyond knowing whether they are above or below the main diagonal. The *Evans-Hoenig test* (also known as *diagonally projected*; Evans and Hoenig 1998) tests for differences in values pooled from off-diagonals that are the same "distance" from the main diagonal (Figure 4.4). In other words, it specifically compares the number of individuals p units below the main diagonal (i.e., cells where $j - i = -p$) to the number of individuals p units above the main diagonal (i.e., cells where $j - i = +p$). The *Bowker test* (also known as *unpooled*; Bowker 1948; Hoenig et al. 1995) tests for differences between cells that are in the same relative positions above and below the main diagonal. In other words, it specifically compares the values in the (i, j)th and (j, i)th cells (Figure 4.4). Formulas for computing these three tests are found in Evans and Hoenig (1998).

Calculating all three tests of symmetry is a two-step process that begins with `ageBias()` as described previously for the age bias plot.[4] Results from the symmetry tests are displayed by submitting the `ageBias()` object to `summary()` along with `what="symmetry"`. Alternatively, use `what="McNemar"`, `what="EvansHoenig"`, or `what="Bowker"` to show results from the individual symmetry tests.

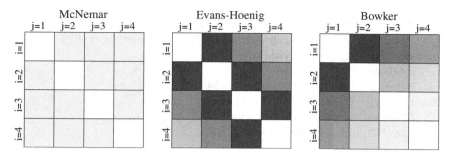

FIGURE 4.4. Schematic age-agreement tables with four ages to demonstrate the calculations for the McNemar, Evans-Hoenig, and Bowker tests of symmetry. Cells or pools of cells in each table that are the same color but on opposite sides of the main diagonal (white cells) are compared in the calculation of the test statistic. Concept modified from McBride et al. (2014).

```
> summary(ab.tA1,what="symmetry")
        symTest df   chi.sq          p
1       McNemar  1 1.580645 0.2086678
2 EvansHoenig  3 2.636364 0.4511502
3        Bowker 10 8.333333 0.5963127
```

The large p-values for all three symmetry tests suggest that the age estimates of Ager A were not systematically biased relative to the true ages.

The age-agreement table, with `refvar` forming the columns, is displayed by submitting the `ageBias()` object to `summary()` with `what="table"`.[5]

```
> summary(ab.tA1,what="table")
        True Age
Ager A 2 3 4 5 6 7 8
     2 - 1 - - - - -
     3 - 6 2 - 1 - -
     4 - 2 3 3 - - -
     5 - - 3 5 3 3 -
     6 - - 1 1 4 3 1
     7 - - - 1 - 2 2
     8 - - - - 2 2 -
```

To partially illustrate the range of results that may appear in age-agreement tables and tests of symmetry, the estimated ages from Agers B and C were also compared to the true ages. Ager B appeared to overestimate the age of young fish and underestimate the age of older fish (Figure 4.5-Left), resulting in a significantly asymmetric age-agreement table, but only as detected by the Bowker test.

```
> ab.tB1 <- ageBias(agerB1~trueAge,data=shad,
                    ref.lab="True Age",nref.lab="Ager B")
> summary(ab.tB1,what="symmetry")
      symTest df    chi.sq          p
1      McNemar  1  0.800000 0.3710934
2 EvansHoenig  2  1.866667 0.3932407
3       Bowker  7 14.800000 0.0386503
> plot(ab.tB1,col.CIsig="black")
```

In contrast, Ager C appeared to underestimate the age of all fish after age 3 (Figure 4.5-Right), resulting in a significantly asymmetric age-agreement table as detected by all three symmetry tests.

```
> ab.tC1 <- ageBias(agerC1~trueAge,data=shad,
                    ref.lab="True Age",nref.lab="Ager C")
> summary(ab.tC1,what="symmetry")
      symTest df   chi.sq           p
1      McNemar  1 30.42222 3.475241e-08
2 EvansHoenig  4 31.78261 2.119129e-06
3       Bowker 16 35.66667 3.218018e-03
> plot(ab.tC1,col.CIsig="black")
```

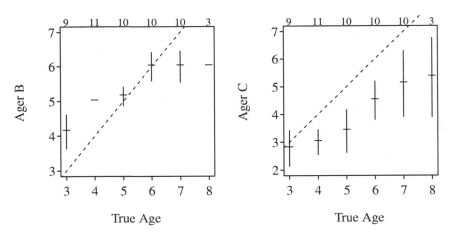

FIGURE 4.5. Age bias plots for comparison of estimated ages from Agers B (Left) and C (Right) to the true ages of American Shad.

4.4 Precision Metrics

There are three commonly used measures of precision among multiple age estimates made on the same fish. *Percent agreement* is the percentage of age estimates for the same fish that either perfectly agree or agree within a certain amount (e.g., within two years of each other). Percent agreement is easy to understand, but may be difficult to interpret because "good agreement" cannot be defined without considering the range of observed ages (Beamish and Fournier 1981). For example, 90% agreement between two readers would be interpreted differently for Coho Salmon with only a few year-classes and Lake Trout with dozens of year-classes. For this reason, percent agreement should not be used as the sole measure of precision among sets of ages.

Beamish and Fournier (1981) proposed the *average percent error* (APE) to measure precision. The APE is defined as

$$APE = 100 * \frac{\sum_{j=1}^{n} \sum_{i=1}^{R} \frac{|x_{ij} - \bar{x}_j|}{\bar{x}_j}}{nR}$$

where x_{ij} is the ith age for the jth fish, $\bar{x}_j$ is the mean age for the jth fish, R is the number of times that each fish was aged (assumed to be the same for all fish), and n is the number of aged fish in the sample.

Chang (1982) suggested that precision should be measured by the *mean coefficient of variation* to avoid the inherent assumption in the APE that the standard deviation of age is proportional to the mean age for individual fish. This measure has been called a CV in the literature, but it is subtly different than a traditional CV (i.e., it is the mean of many CVs) and, thus, will be called *average CV* (ACV) hereafter. The ACV is defined as

$$ACV = 100 * \frac{\sum_{j=1}^{n} \frac{s_j}{\bar{x}_j}}{n}$$

where s_j is the standard deviation of the R age estimates for the jth fish.

All three measures of precision are calculated in a two-step process that begins with `agePrecision()` from **FSA**. The first argument to `agePrecision()` is a formula of the form `~var1+var2+...`, where `var1`, `var2`, etc. are the variables that contain the multiple age estimates.[6] The data.frame that contains the variables must be given in `data=`.

```
> ap.A <- agePrecision(~agerA1+agerA2,data=shad)
```

Two types of percentage agreement tables may be extracted by submitting the `agePrecision()` object to `summary()`. The percentage table of raw differences is extracted with `what="difference"`.

```
> summary(ap.A,what="difference")
    -2     -1      0      1      2      3
 3.922 15.686 50.980 19.608  7.843  1.961
```

These results show that both estimates by Ager A agreed for 51.0% of the fish
and that the second estimate was two years older for 3.9% of the fish, but two
years younger for 7.8% of the fish.

Use `what="absolute difference"` to extract the percentage table of the
absolute value of the differences.

```
> summary(ap.A,what="absolute difference")
     0      1      2      3
50.980 35.294 11.765  1.961
```

These results show that the two age estimates agreed for 51.0% of the fish (as
above) and differed by two years (either way) for 11.8% of the fish.

The mean APE and ACV are extracted by including `what="precision"`
in `summary()`.

```
> summary(ap.A,what="precision")
  n R    ACV   APE PercAgree
 53 2 9.739 6.887     52.83
```

These results show a general lack of precision between the two estimates by
Ager A as the mean ACV (9.74%) exceeds the 5% level suggested by Campana
(2001) to represent precise estimates (see Section 4.5).

The age precision results may be computed for more than two age estimates. For example, the results below examine precision among the first
readings made by all three agers.[7]

```
> ap.ABC <- agePrecision(~agerA1+agerB1+agerC1,data=shad)
> summary(ap.ABC,what="difference")
                  -2     -1      0      1      2      3      4
agerA1 - agerB1 18.182 15.152 45.455 12.121  9.091  0.000  0.000
agerA1 - agerC1  0.000  5.882 19.608 41.176 19.608 11.765  1.961
agerB1 - agerC1  0.000  6.061  6.061 36.364 36.364 15.152  0.000
> summary(ap.ABC,what="precision")
  n R    ACV  APE PercAgree
 53 3 22.98 16.7     41.51
```

4.5   Further Considerations

An alternative to the age bias plot introduced in Section 4.2 is the *Bland-Altman* (Altman and Bland (1983) and numerous subsequent citations) or *Tukey mean-difference* plot. The Bland-Altman plot is popular in the chemistry and medical fields where it is often used to analyze agreement between two measurements of the same quantity (e.g., assays). Versions of the Bland-Altman plot are constructed with `BA.plot()` from **MethComp** (Carstensen et al. 2015) and `bland.altman.plot()` from **BlandAltmanLeh** (Lehnert 2014). The Tukey mean-difference plot may be constructed with `tmd()` from **lattice**.

The three tests of symmetry introduced in Section 4.3 behave differently depending on sample size and the pattern of asymmetry present (Evans and Hoenig 1998; Kimura and Anderl 2005; McBride 2015). A consensus recommendation on which test of symmetry to use for age data has not yet emerged in the literature. However, McBride (2015) suggested, from a limited suite of simulations, that the McNemar test is generally not needed and that the Evans-Hoenig test performed well under most conditions simulated, though the Bowker test performed equally well in many simulations. However, the Evans-Hoenig test may be more capable of detecting asymmetries with smaller sample sizes (Evans and Hoenig 1998; Kimura and Anderl 2005). The ultimate choice for which method to use may depend on the type of symmetry one wants to detect.

The ACV is greater than the APE by an average of 41% (Campana 2001) or a factor of $\sqrt{2}$ (Kimura and Anderl 2005) when $R = 2$. Thus, the two measures are functionally interchangeable (McBride 2015) and it does not matter whether ACV or APE is used for relative comparisons. However, ACV may be more familiar and easier to interpret (Kimura and Anderl 2005). Global benchmarks for diagnosing acceptable levels of precision have not been proposed because precision is influenced by the species and the structure used to assign age (Campana 2001). Campana (2001) did suggest that an ACV of less than 5% suggests that the assigned ages for fishes of moderate longevity and structure reading complexity are acceptably precise.[8]

Serious comparisons of multiple sets of ages should use a combination of the methods described in this chapter. McBride (2015, p. 14) summarized this thought as follows

> ... [N]o single approach seems sufficient to describe or interpret ageing error. A strategic complement of approaches should: (i) qualitatively evaluate raw data plots for bias and precision, (ii) check that precision levels support acceptable repeatability and are suitable to perform a test of symmetry, and (iii) test the symmetry of the data for biases evident in the qualitative evaluation. Combined, these approaches compensate for the weaknesses inherent in any single approach.

Notes

[1] A variety of methods for adjusting p-values for multiple comparisons may be used. The default is to use the "Holm" method; type `p.adjust.methods()` to see the full list of possible adjustment methods that may be used in `method=` in `ageBias()`.

[2] As mentioned previously, the confidence intervals in the age bias plot are unadjusted for multiple comparisons. The p-value in the output from `summary()`, however, is adjusted for multiple comparisons. Unadjusted p-values are returned by `summary()` if `method="none"` is used in the initial `ageBias()` call.

[3] Historically, systematic bias between two sets of age estimates has also been examined with simple linear regression, matched-pairs t-test, and Wilcoxon rank test. Campana et al. (1995) showed that these three tests have deficiencies in several situations commonly encountered with fisheries age data and, thus, none of these methods are commonly used in modern analyses. Simple linear regression is described in Section 7.3. A matched-pairs t-test is conducted with `t.test()` using `paired=TRUE`. The Wilcoxon test is conducted with `wilcox.test()` using `paired=TRUE`.

[4] The Bowker and the Evans-Hoeing tests of symmetry can also be computed with `compare2()` from **fishmethods** (Nelson 2015). The `plotagecomparisons()` function (available from `http://fluke.vims.edu/hoenig/useful_functions_in_R.htm`) also provides a novel method for visually comparing three age estimates.

[5] By default, the zeroes in the table are replaced with -. This behavior may be changed by setting `zero.print=` to a different string.

[6] If only two variables are being considered, then the formula to `agePrecision()` may be the same as that used in `ageBias()`.

[7] The sample size is much smaller in the example where the three agers are compared because Ager C did not estimate an age for several fish.

[8] Of course, this benchmark value could be deflated for use with APE; that is, use $\frac{5}{\sqrt{2}} = 3.5$.

5

Age-Length Keys

As described in Chapter 4, the ages of fish are important data for understanding the dynamics of fish populations. Estimating age for a large number of fish, however, is labor intensive. Fortunately, there is usually a strong relationship between length and age, and measuring the length of a large number of fish is relatively easy. Thus, the age structure for a large sample of fish can be reliably[1] estimated by summarizing the relationship between age and length for a subsample of fish and then applying this summary to the entire sample. The summary of the relationship between age and length is an *age-length key* (ALK). This chapter will illustrate methods to construct an ALK and how to apply that ALK to estimate the age structure of the larger sample of fish.

Required Packages for This Chapter

Functions used in this chapter require loading the packages shown below.[2]

```
> library(FSA)
> library(magrittr)
> library(dplyr)
> library(nnet)
```

5.1 Foundational Background

Suppose that a random sample of N fish are captured with the goal of determining the age structure of those fish. If N is not too large, then age may be estimated for all N fish. However, if N is large, then a subsample of n fish from the entire range of lengths present in the N fish is selected for age estimation. This latter situation is an example of *two-stage sampling*; that is, one sample of N fish is taken from the extant population and a second (sub)sample of n fish is taken from the original sample. Two-stage sampling for age estimation is more common than random sampling and, thus, is the focus of the rest of this chapter.

In two-stage sampling, the n fish may be completely randomly selected from the N fish, but it is more common to randomly select fish *from within each of L length intervals.* The number of fish sampled for aging in each length interval (n_i) can either be fixed at a constant number or proportional to the total number of fish in that length interval (N_i). Proportional selection produces statistically "better" results (Kimura 1977), but proportional selection is not often used in the field because the length of all fish would need to be known before taking the subsample.

The subsample of n fish for which length and age is recorded is called the *aged sample.* In addition to length and age, the length interval to which the fish belongs is recorded.[3] The numbers of fish in the ith length interval and of the jth age (n_{ij}) are summarized in a frequency table. An example frequency table is shown below (with generic labels for the ages and length intervals).

	A_1	A_2
L_1	6	2
L_2	3	3
L_3	1	4

In this example, six fish in the aged sample belong to the first length interval and first age (i.e., $n_{11}=6$). Similarly, one fish belongs to the third length interval and first age (n_{31}).

An age for one of the $N - n$ fish for which age was not estimated will be based on the proportions of each age within the fish's length interval in the aged sample. These conditional proportions are derived from the aged sample by dividing each cell of the frequency table by the total number of fish in that length interval (i.e., each cell is divided by the sum of its row). Thus, the conditional proportion of fish in the ith length interval that are age j is $p_{j|i} = \frac{n_{ij}}{n_i}$. The table below shows the conditional proportions derived from the frequency table shown above.

	A_1	A_2
L_1	0.75	0.25
L_2	0.50	0.50
L_3	0.20	0.80

Thus, for example, the proportion of fish in the first length interval that are the first age is $p_{1|1}=0.75$ and the proportion of fish in the third length interval that are the first age is $p_{1|3}=0.20$. The table of conditional probabilities derived from the aged sample is the ALK.

The proportions in the ALK serve as probabilities for assigning ages to the fish that were not part of the aged sample. For example, the probability that an unaged fish with a length in the first length interval is the first age is 0.75. In other words, one would expect 75% of the unaged fish with a length in the first length interval to be the first age.

5.2 Constructing an Age-Length Key

5.2.1 Preparing the Aged Sample

In most situations, a single data file contains the lengths of all N sampled fish with corresponding ages for the n fish in the aged (sub)sample and missing values (i.e., as NA) for the $N - n$ fish for which ages were not estimated. As an example, the total lengths (nearest mm) and ages (from scales) of Creek Chub were recorded in *CreekChub.csv* (from Quist et al. 2012). An examination of the first and last three rows of the data.frame shows fish with and without estimated ages.

```
> cc <- read.csv("CreekChub.csv")
> headtail(cc)
    len age
1    41   0
2    42   0
3    42   0
216 109  NA
217 109  NA
218 109  NA
```

The appropriate length intervals for each fish are appended to the data.frame with lencat() from FSA as described in Section 2.2.6. However, if a length interval in the aged sample contains no fish, then you may want to use as.fact=TRUE and drop.levels=FALSE in lencat() to make sure that length intervals with no data exist in the ALK matrix used in Section 5.2.2 and the plots in Section 5.3. However, the smoothed ALK in Section 5.2.3 requires the nonfactored version of the length categorization variable (i.e., the default from lencat()).

For all analyses below, the appropriate 10 mm wide length interval for each Creek Chub is stored in a new variable called lcat10.

```
> cc %<>% mutate(lcat10=lencat(len,w=10))
> headtail(cc)
    len age lcat10
1    41   0     40
2    42   0     40
3    42   0     40
216 109  NA    100
217 109  NA    100
218 109  NA    100
```

The original data.frame with all N fish must be separated into one data.frame that contains the n fish in the aged sample and a second data.frame that contains the $N - n$ fish not in the aged sample. The is.na() function, which takes a vector of data as its only argument, is used to identify positions in a vector that are NA. The code below demonstrates that is.na() returns FALSE for the first three rows of the cc data.frame because those fish had age estimates. In contrast, TRUE is returned for the last three rows because those fish did not have ages (i.e., were NA).

```
> is.na(headtail(cc)$age)          # demonstration purposes only
[1] FALSE FALSE FALSE  TRUE  TRUE  TRUE
```

With this, the unaged sample is extracted by retaining all rows where is.na() returns TRUE. The aged sample, however, is obtained by extracting all rows where the age variable is "not NA" or, in code, where !is.na() returns TRUE (the ! returns the complement of a logical vector).

```
> cc.unaged <- filter(cc,is.na(age))
> headtail(cc.unaged)
    len age lcat10
1    45  NA     40
2    50  NA     50
3    51  NA     50
113 109  NA    100
114 109  NA    100
115 109  NA    100
> cc.aged <- filter(cc,!is.na(age))
> headtail(cc.aged)
    len age lcat10
1    41   0     40
2    42   0     40
3    42   0     40
101 176   4    170
102 198   4    190
103 205   4    200
```

The unaged sample should have all NA values, whereas the aged sample should not have any NA values for the age variable. The all() and any() functions may be used to check that these data.frames contain the expected fish. Specifically, all() returns TRUE if *all* items in a logical vector are TRUE, whereas any() returns TRUE if *any* item in a logical vector is TRUE.

```
> all(is.na(cc.unaged$age))        # better be TRUE
[1] TRUE
> any(is.na(cc.aged$age))          # better be FALSE
[1] FALSE
```

The aged subsample is used to construct the ALKs in Sections 5.2.2 and 5.2.3. The unaged sample is summarized for applications in Section 5.4 and used more explicitly in Section 5.4.4.

5.2.2 Observed Age-Length Keys

The ALK is constructed from the aged sample in two steps,[4] similar to how the ALK was introduced in Section 5.1. First, the frequency of fish in each length interval and age combination is constructed with xtabs(). The first argument to xtabs() is a formula of the form ~lcat+age, where lcat contains the length intervals and age contains the estimated ages. The name of the data.frame that contains these variables is given in data=.

```
> ( alk.freq <- xtabs(~lcat10+age,data=cc.aged) )
       age
lcat10  0  1  2  3  4
    40 10  0  0  0  0
    50  3  7  0  0  0
    60  0 10  0  0  0
    70  0 10  0  0  0
    80  0  8  0  0  0
    90  0  8  2  0  0
   100  0  5  5  0  0
   110  0  0 10  0  0
   120  0  0  6  0  0
   130  0  0  3  0  0
   140  0  0  1  5  0
   150  0  0  1  1  0
   160  0  0  1  1  1
   170  0  0  0  1  2
   190  0  0  0  0  1
   200  0  0  0  0  1
```

While it is not a necessary calculation, computing the sum of the rows in the frequency table (obtained by submitting the xtabs() object to rowSums()) gives the number of aged fish in each length interval. This summary may indicate how the aged fish were (sub)sampled. For example, the row sums for the Creek Chub data corroborate that a fixed sample of as many as 10 fish per 10 mm length interval were sampled for age estimation.

```
> rowSums(alk.freq)
 40  50  60  70  80  90 100 110 120 130 140 150 160 170 190 200
 10  10  10  10   8  10  10  10   6   3   6   2   3   3   1   1
```

The conditional proportions that form the ALK are calculated by dividing each cell of the frequency table by the sum of the corresponding row.

These row proportions are constructed by submitting the `xtabs()` object to `prop.table()` and including `margin=1` to indicate that the proportions are computed by row.

```
> alk <- prop.table(alk.freq,margin=1)
> round(alk,3)      # rounded for display purposes only
      age
lcat10     0     1     2     3     4
    40 1.000 0.000 0.000 0.000 0.000
    50 0.300 0.700 0.000 0.000 0.000
    60 0.000 1.000 0.000 0.000 0.000
    70 0.000 1.000 0.000 0.000 0.000
    80 0.000 1.000 0.000 0.000 0.000
    90 0.000 0.800 0.200 0.000 0.000
   100 0.000 0.500 0.500 0.000 0.000
   110 0.000 0.000 1.000 0.000 0.000
   120 0.000 0.000 1.000 0.000 0.000
   130 0.000 0.000 1.000 0.000 0.000
   140 0.000 0.000 0.167 0.833 0.000
   150 0.000 0.000 0.500 0.500 0.000
   160 0.000 0.000 0.333 0.333 0.333
   170 0.000 0.000 0.000 0.333 0.667
   190 0.000 0.000 0.000 0.000 1.000
   200 0.000 0.000 0.000 0.000 1.000
```

Thus, for example, 30.0% of Creek Chubs in the 50 mm length interval were age-0 and 70.0% were age-1. Note that no fish in the 180 mm length interval were present in the aged sample.

5.2.3   Smoothed or Modeled Age-Length Keys

Two issues commonly arise with an observed ALK. First, highly variable ages within a length interval and small sample sizes in some length intervals may result in portions of an ALK that are biologically "weird." For example, a higher proportion of younger than older fish may be present in a length interval of relatively large fish (which is exacerbated if the adjacent length intervals have lower proportions for the younger age). There is some evidence for this phenomenon for age-2 Creek Chubs in the 150 mm length interval. Second, in some instances it may not be possible to ensure that fish are aged from each sampled length interval (as would occur, for example, if sampling continued after the aged sample had been taken). In these instances, the ALK will have a missing length interval and, thus, no information about the age distribution would exist that could be used to assign ages to unaged fish in that interval.

These two issues may be addressed with an ALK created from $p_{j|i}$ values predicted by statistical models fit to the observed length interval and age data (Gerritsen et al. 2006). These models are fit to *all length intervals and ages*; thus, the predicted proportion of fish at age for any one length interval is influenced not only by the data for that length interval and age, but also by the data for other length intervals and ages. This characteristic of these models results in predicted values that follow a smooth curve, as compared to the more variable representations in the observed ALK. Additionally, the $p_{j|i}$ values can be predicted from these models for length intervals for which no fish were aged. Thus, length intervals with no aged fish can be populated with proportions.

One example of such a model is the multinomial logistic regression model (Gerritsen et al. 2006). A multinomial logistic regression model is similar to the binomial logistic regression model (commonly used, for example, to model maturity by length or age), except that the response variable has more than two possible levels (i.e., estimated age usually has more than two levels). The multinomial logistic regression model is fit with **multinom()** from **nnet** (Venables and Ripley 2002). The first argument is a formula of the form **age~lcat**, where **age** contains the estimated ages and **lcat** contains the length intervals. The data.frame with these variables for the aged sample is given in **data=**. The **multinom()** function uses an iterative algorithm that will stop, by default, after 100 iterations. If the algorithm reaches 100 iterations before reaching a solution, then it is prudent to raise the maximum number of iterations with **maxit=**, as illustrated below (the iteration information was suppressed from the output).

```
> cc.mlr <- multinom(age~lcat10,data=cc.aged,maxit=500)
```

Predictions from the fitted model are made with **predict()**. The first argument is the **multinom()** object. The second argument is a data.frame that contains values for the explanatory variable (length intervals, in this case) at which the probabilities of each age should be predicted. This data.frame must contain a variable that is named exactly as the **lcat** variable in the **multinom()** call. The length intervals for which to make the predictions are best constructed with **seq()**, with the minimum length interval value, the maximum length interval value, and the step value for the sequence as the three arguments. Finally, include **type="probs"** so that probabilities, rather than logit odds, are returned by **predict()**.

```
> lens <- seq(40,200,10)
> alk.sm <- predict(cc.mlr,data.frame(lcat10=lens),type="probs")
> row.names(alk.sm) <- lens    # for clarity
> round(alk.sm,3)                  # round for display purposes only
         0     1     2     3     4
40   0.997 0.003 0.000 0.000 0.000
50   0.301 0.699 0.000 0.000 0.000
60   0.000 0.999 0.000 0.000 0.000
```

```
70   0.000 0.999 0.001 0.000 0.000
80   0.000 0.987 0.013 0.000 0.000
90   0.000 0.874 0.126 0.000 0.000
100  0.000 0.382 0.614 0.004 0.000
110  0.000 0.052 0.929 0.019 0.000
120  0.000 0.005 0.931 0.064 0.000
130  0.000 0.000 0.812 0.187 0.000
140  0.000 0.000 0.562 0.434 0.004
150  0.000 0.000 0.267 0.693 0.040
160  0.000 0.000 0.077 0.667 0.256
170  0.000 0.000 0.009 0.278 0.713
180  0.000 0.000 0.001 0.055 0.944
190  0.000 0.000 0.000 0.009 0.991
200  0.000 0.000 0.000 0.001 0.999
```

Thus, for example, 30.1% of Creek Chubs in the 50 mm length interval are predicted to be age-0 and 69.9% are predicted to be age-1. Note that in this modeled ALK (as compared to the observed ALK from the previous section) that no proportion is exactly 1 and that proportions exist for the 180 mm length interval for which no fish existed in the aged sample.

5.3   Visualizing an Age-Length Key

The ALK, whether observed or smoothed, may be displayed as a series of stacked bars, stacked areas, or connected lines over length interval labels, with different colors or shadings for each age (Figure 5.1). These plots are constructed with alkPlot() from **FSA**, with the ALK from prop.table() as the first argument. A stacked bar plot is the default. Stacked area and line plots are constructed by including type="area" or type="line", respectively, in alkPlot(). The ages are denoted by a set of richly-contrasting colors by default, but this may be changed to shades of gray with pal="gray". Finally, a legend for the colors is added to the area plot with showLegend=TRUE. The size of the legend may be decreased or increased from the default size by including a number less or more than 1, respectively, in leg.cex=.[5]

```
> alkPlot(alk,type="area",pal="gray",showLegend=TRUE,
          leg.cex=0.7,xlab="Total Length (mm)")
```

The ALK is also commonly shown as a "bubble plot," where circles with areas proportional to the number of fish in each length interval that are a given age are placed at the coordinates of the length intervals and ages (Figure 5.2).

```
> alkPlot(alk,type="bubble",xlab="Total Length (mm)")
```

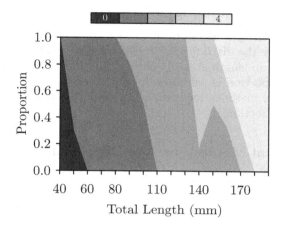

FIGURE 5.1. Area plot representation of the observed age-length key for Creek Chubs.

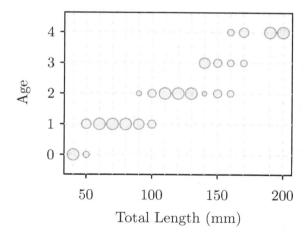

FIGURE 5.2. Bubble plot representation of the observed age-length key for Creek Chubs. The area of each circle is proportional to the proportion of fish in a length interval that are a given age.

5.4   Apply an Age-Length Key

Age-length keys are typically applied to the complete sample of fish with the goal of estimating the distribution of ages or some other biological value for each age (e.g., mean length-at-age). The distribution of ages may be described by either the proportion of fish at each age (p_j) or the number of fish at each age (N_j). Various methods for computing these two types of summaries are described in this section.

5.4.1   Classical Method for Age Distribution

In its simplest form, the proportion of fish in the ith length interval to be assigned the jth age is equal to the proportion of all N fish that are in the ith length interval (l_i) times the conditional probability that a fish in the ith length interval is age j; that is, $p_{ij} = l_i p_{j|i}$. The proportion of N fish to be assigned age j is then

$$p_j = \sum_{i=1}^{L} p_{ij} = \sum_{i=1}^{L} l_i p_{j|i}$$

Returning to the simple hypothetical example from Section 5.1, if 50% of the entire sample of fish is in the first length interval, then the proportion of fish in the first length interval to be assigned the first age is 0.375 (=0.50*0.75). If 30% and 20% of the entire sample is in the other two length intervals, then the proportion of all fish to be assigned the first age is 0.565 (=0.50*0.75+0.30*0.50+0.20*0.20).

The number of fish to be assigned the jth age is the total number of fish times the proportion to be assigned the jth age; that is, $N_j = Np_j$. Thus, for example, if there are 60 fish in the entire sample, then 33.9 (=60*0.565) are of the first age. The calculation of N_j often results in decimals. Rounding the result removes the decimals, but may result in one fewer or more fish in some situations.

These calculations are more easily performed in R by first computing the N_j values and then dividing each of these by N to compute the p_j values. The frequency of *all* fish in each length interval (i.e., the N_i values) is found with `xtabs()`.

```
> ( len.n <- xtabs(~lcat10,data=cc) )
lcat10
 40  50  60  70  80  90 100 110 120 130 140 150 160 170 190 200
 11  33  42  32   8  25  32  10   6   3   6   2   3   3   1   1
```

The `sweep()` function is used to multiply each row in the ALK matrix (i.e,. the $p_{j|i}$) by the corresponding position in the vector of N_i values. The ALK matrix is the first argument to `sweep()`. Additionally, `FUN="*"` tells `sweep()` to multiply rows (i.e., `MARGIN=1`) of the ALK by the corresponding element in the vector given in `STATS=` (which, in this case, is the vector of N_i values). The result is a matrix of all fish allocated by length interval and age.

```
> ( tmp <- sweep(alk,MARGIN=1,FUN="*",STATS=len.n) )
     age
lcat10    0    1     2    3    4
   40  11.0  0.0   0.0  0.0  0.0
   50   9.9 23.1   0.0  0.0  0.0
   60   0.0 42.0   0.0  0.0  0.0
   70   0.0 32.0   0.0  0.0  0.0
   80   0.0  8.0   0.0  0.0  0.0
   90   0.0 20.0   5.0  0.0  0.0
  100   0.0 16.0  16.0  0.0  0.0
  110   0.0  0.0  10.0  0.0  0.0
  120   0.0  0.0   6.0  0.0  0.0
  130   0.0  0.0   3.0  0.0  0.0
  140   0.0  0.0   1.0  5.0  0.0
  150   0.0  0.0   1.0  1.0  0.0
  160   0.0  0.0   1.0  1.0  1.0
  170   0.0  0.0   0.0  1.0  2.0
  190   0.0  0.0   0.0  0.0  1.0
  200   0.0  0.0   0.0  0.0  1.0
```

The number of fish allocated to each age (N_j) is obtained from this matrix by summing the columns with `colSums()`.[6]

```
> ( ad1 <- colSums(tmp) )
    0     1     2     3     4
 20.9 141.1  43.0   8.0   5.0
```

Submitting these column sums to `prop.table()` gives the proportion of fish at each age (p_j).

```
> round(prop.table(ad1),3)   # rounded for display purposes only
    0     1     2     3     4
0.096 0.647 0.197 0.037 0.023
```

Thus, for example, approximately 141 Creek Chubs or 64.7% of the complete sample are estimated to be age-1.

5.4.2 Age Distribution with Standard Errors

Quinn II and Deriso (1999) showed that, with two-stage sampling, the standard error of the proportion of fish at each age is[7]

$$SE(p_j) = \sqrt{\sum_{i=1}^{L} \frac{l_i^2 p_{j|i}(1 - p_{j|i})}{n_i - 1} + \frac{l_j(p_{j|i} - p_i)^2}{N}}$$

The SE of the number of fish at each age is then $SE(N_j) = NSE(p_j)$.

The p_j and $SE(p_j)$ are computed with `alkAgeDist()` from **FSA**.[8] The first argument is the ALK matrix. Vectors of sample sizes in each length interval for the aged and overall samples are given in `lenA.n=` and `len.n=`, respectively.

```
> # note: alk.freq and len.n were calculated previously
> alkAgeDist(alk,lenA.n=rowSums(alk.freq),len.n=len.n)
  age       prop         se
1   0 0.09587156 0.028037207
2   1 0.64724771 0.044703252
3   2 0.19724771 0.036408292
4   3 0.03669725 0.013789774
5   4 0.02293578 0.008754956
```

Again, approximately 64.7%, with a SE of 4.5%, of the complete sample are estimated to be age-1.

5.4.3   Mean Length-at-Age

The mean (and standard deviation) length-at-age computed from just the fish in the aged sample is biased if the aged sample was obtained by sampling a fixed number of fish per length interval. Bettoli and Miranda (2001) show, however, that an unbiased estimate of the mean length-at-age ($\bar{L}_j$) for the N fish is

$$\bar{L}_j = \frac{N}{N_j} \sum_{i=1}^{L} p_{ij} \bar{L}_{ij}$$

where $\bar{L}_{ij}$ is the mean length of fish from the aged sample in the ith length interval that were age j. The corresponding standard deviation for length-at-age is

$$SD(L_j) = \sqrt{\frac{N}{N_j - 1} \sum_{i=1}^{L} p_{ij}(\bar{L}_{ij} - \bar{L}_j)^2}$$

The $\bar{L}_{ij}$ and $\bar{L}_j$ could be replaced with any other quantitative variable (e.g., weight) measured on each individual.

These calculations are made with `alkMeanVar()` from FSA.[9] The first argument is the ALK matrix. The second argument is a formula of the form `var~lcat+age`, where `var` contains the quantitative variable of interest (e.g., length, weight), `lcat` contains the length interval for each fish (must be the same length intervals used to construct the ALK), and `age` contains the age of each fish. The data.frame that contains these variables for the *aged* sample is given in `data=`. The vector of sample sizes in each length interval for the *overall* sample is given in `len.n=`.

```
> alkMeanVar(alk,len~lcat10+age,data=cc.aged,len.n=len.n)
  age     mean       sd
1   0 49.05263  5.458119
2   1 74.07725 17.232147
3   2 112.98372 16.054776
4   3 151.87500 10.960155
5   4 183.20000 17.163187
```

Thus, the estimated mean length of age-1 Creek Chubs is 74.1 mm with a standard deviation of 17.2 mm.

5.4.4 Assign Individual Ages and Summarize

Isermann and Knight (2005) described a method where individual fish in the unaged sample are assigned an age based on an ALK. Thus, all N fish will then have a recorded age. Isermann and Knight (2005) noted that the advantage of this method is that standard statistical formulae can be used to estimate variability for measures like mean length-at-age and standard statistical methods (e.g., t-tests, analysis of variance, likelihood ratio tests) can be used to compare age-based summaries among groups (e.g., compare mean lengths-at-age between sexes).

In the Isermann-Knight method, the expected number of unaged fish in the ith length interval to be assigned the jth age is initially computed with $N_{ij}^* = N_i^* p_{j|i}$ where the N_i^* are derived from the *UNAGED FISH ONLY*. Again, returning to the simple hypothetical example from Section 5.1 with 16 unaged fish in the first length interval, twelve (i.e., $16 * 0.75$) fish would be assigned the first age and four (i.e., $16 * 0.25$) would be assigned the second age. Which of the 16 fish would be assigned the specific ages is randomly determined, but the number of fish to be assigned each age is set at twelve and four, respectively.

One difficulty that soon becomes apparent is when the expected N_{ij}^* is not a whole number. For example, suppose that 15 unaged fish are in the second length interval. In this case, the expected number of fish for each age is 7.5 (i.e., $15 * 0.5$). Clearly, half a fish cannot be assigned to a given age. This difficulty was called *fractionality* by Isermann and Knight (2005).

Fractionality is handled by initially reducing N_{ij}^* to the closest integer *smaller* than the expected N_{ij}^*. In the example above, seven fish would be assigned each of the two ages. This allocation, however, leaves one fish without an assigned age. The age for this fish is randomly assigned to one of the ages with the randomization based on a probabilistic weight equal to the corresponding $p_{j|i}$. In this example, the "extra" fish would be assigned the first age with a probability of 0.50 and would be assigned the second age with a probability of 0.50. If the randomization process determined that the "extra" fish should be in the second age, then seven fish would be assigned the first age and eight would be assigned the second age. Again, the specific age to be assigned to each fish is randomly determined, but seven fish would be assigned the first age and eight fish would be assigned the second age as determined above.

The N_{ij}^* are determined, fractionality is handled, and the random ages are assigned to unaged fish according to the Isermann-Knight method with `alkIndivAge()` from **FSA**. The first argument is the ALK (from `prop.table()`). The second argument is a formula of the form `age~len`.[10] The data.frame with these variables for fish in the *UNAGED* sample is given in `data=`. The function returns a data.frame that has the same number of rows as the data.frame given in `data=`, but each fish will have an age that was assigned according to the Isermann-Knight method. The results of `alkIndivAge()` should be assigned to an object.

```
> cc.unaged.mod <- alkIndivAge(alk,age~len,data=cc.unaged)
> head(cc.unaged.mod)
  len age lcat10
1  45   0     40
2  50   1     50
3  51   1     50
4  51   1     50
5  52   1     50
6  52   1     50
```

The new data.frame with assigned ages may be row-bound (using `rbind()`) with the data.frame that contains the aged sample to form a data.frame of all the original fish, now with estimated ages for all fish.

```
> cc.fnl <- rbind(cc.aged,cc.unaged.mod)
```

The proportional age distribution is then computed from the combined data.frame with `xtabs()` and `prop.table()`.

```
> ( ad3 <- xtabs(~age,data=cc.fnl) )
age
  0   1   2   3   4
 20 142  43   8   5
> round(prop.table(ad3),3)    # rounded for display purposes only
age
    0     1     2     3     4
0.092 0.651 0.197 0.037 0.023
```

Thus, 142 fish or 65.1% of all sampled fish are estimated to be age-1.

The means and standard deviations of lengths-at-age may be computed using group_by() and summarize() from **dplyr** as described in Section 2.5.

```
> cc.sumlen <- cc.fnl %>% group_by(age) %>%
    summarize(n=validn(len),mn=mean(len,na.rm=TRUE),
            sd=sd(len,na.rm=TRUE),se=se(len,na.rm=TRUE)) %>%
    as.data.frame()
> cc.sumlen
  age   n       mn        sd       se
1   0  20  49.45000  6.419338 1.435408
2   1 142  74.40845 16.712061 1.402445
3   2  43 113.83721 15.579184 2.375803
4   3   8 151.87500 11.179541 3.952565
5   4   5 183.20000 17.253985 7.716217
```

Thus, the estimated mean length of age-1 Creek Chubs is 74.4 mm with a standard deviation of 16.7 mm.

Finally, because ages were assigned to individuals, a plot of individual lengths-at-age with mean lengths-at-age superimposed (Figure 5.3) may be constructed with plot() and lines() as described in Section 3.2.2.

```
> plot(len~age,data=cc.fnl,pch=19,col=rgb(0,0,0,1/10),
      xlab="Age",ylab="Total Length (mm)",ylim=c(0,205))
> lines(mn~age,data=cc.sumlen,lwd=2,lty=2)
```

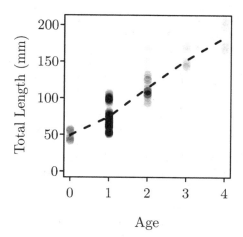

FIGURE 5.3. Length-at-age with mean length-at-age (dashed line) superimposed for the complete sample of Creek Chub.

5.5   Among Group Statistical Comparisons

Separate age-length keys may be needed for different groups of fish (e.g., by sex
or location) if the relationship between age and length differs among groups.
The multinomial model introduced in Section 5.2.3 can be extended to provide
a statistical test for whether two or more ALKs differ among groups (Gerritsen
et al. 2006). Specifically, the test described below is used to determine if the
distribution of lengths within each age is similar or not among the groups.

A likelihood ratio test to identify a difference in ALKs between two groups
requires the fitting of two multinomial models. The first simpler model has
length interval as the only explanatory variable (i.e., the model from Section
5.2.3). The second more complex model has length interval, the factor variable
that identifies the groups, and the interaction between these two variables as
explanatory variables. The p-value for testing the effect of the group (i.e., the
factor variable) in explaining the distribution of lengths within each age is
obtained by computing a chi-square test-statistic as the difference in -2*log
likelihood values for the two models. The degrees-of-freedom for this test are
equal to the difference in degrees-of-freedom between the two models.

As an example, ages were estimated from otolith thin sections for a subsample of all Siscowet morphotype Lake Trout collected from Lake Superior near
Grand Marais, Michigan. The data were recorded in *SiscowetMI2004.csv*.
Data preparation included removing all fish for which sex was not determined,
isolating the aged fish, and creating a variable with the 25 mm length interval
for each fish.

```
> sis <- read.csv("SiscowetMI2004.csv") %>%
    filter(!is.na(age),!is.na(sex)) %>%
    mutate(lcat=lencat(len,w=25))
```

The two models required for the likelihood ratio test to determine if the
ALKs differed by sex were fit with `multinom()`. In the second model, note
that the right-hand side of the formula (`lcat*sex`) is short for the long form
of `lcat+sex+sex:lcat`, where the colon denotes an interaction term.

```
> mod1 <- multinom(age~lcat,data=sis,maxit=500)
> mod2 <- multinom(age~lcat*sex,data=sis,maxit=500)
```

The likelihood ratio test is computed with `anova()` using the simpler model
as the first argument and the more complex model as the second argument.

```
> anova(mod1,mod2)
Likelihood ratio tests of Multinomial Models

Response: age
          Model Resid. df Resid. Dev    Test    Df LR stat. Pr(Chi)
1          lcat      2758    677.301
2 lcat * sex         2730    653.868 1 vs 2     28  23.4327 0.711093
```

The large p-value ($p = 0.7111$) indicates that there is no statistical difference in the ALKs between male and female Siscowet Lake Trout. In other words, the distribution of lengths within each age are statistically similar for all ages between male and female Siscowet Lake Trout.

The smoothed ALKs for each sex may be constructed from the more complex model. In this case, data.frames should be constructed that contain the lengths at which to make the predictions *AND* a variable that contains the sex for the predictions.

```
> lens <- seq(350,675,25)
> dfF <- data.frame(lcat=lens,sex="F")
> dfM <- data.frame(lcat=lens,sex="M")
```

The ALKs for each sex are constructed from the predictions for each sex as shown in Section 5.2.3 and the ALKs are visualized using alkPlot() as described in Section 5.3.

```
> alkF <- predict(mod2,dfF,type="probs")
> rownames(alkF) <- lens
> alkPlot(alkF,type="area",pal="gray",xlab="Total Length (mm)",
          showLegend=TRUE,leg.cex=0.7)
>
> alkM <- predict(mod2,dfM,type="probs")
> rownames(alkM) <- lens
> alkPlot(alkM,type="area",pal="gray",xlab="Total Length (mm)",
          showLegend=TRUE,leg.cex=0.7)
```

While these two ALK appear to differ (Figure 5.4), the results of the statistical test above suggest that they do not differ significantly.

5.6 Further Considerations

Due to the importance of age data in fisheries management, the methodology related to age-length keys is ever evolving. The following are entry points

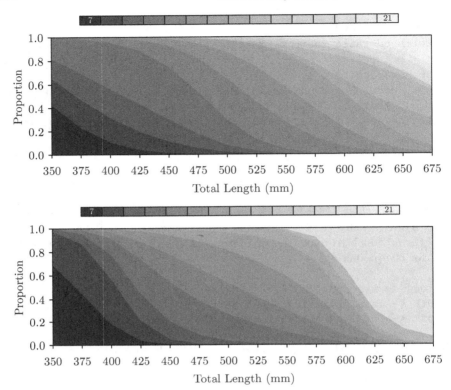

‚FIGURE 5.4. Area plot representations of the smoothed age-length keys for female (top) and male (bottom) Siscowet Lake Trout from Michigan waters of Lake Superior in 2004.

(i.e., not an exhaustive list) to the literature for methods not discussed in this chapter. Kimura and Chikuni (1987), Hoenig and Heisey (1987), Hoenig et al. (1993, 1994), and Gascuel (1994) provided a variety of methods for using other (e.g., past) length frequency information along with the ALK. These methods are implemented in the **ALKr** package. Morton and Bravington (2008) described how the ALK can be augmented with length frequency and growth information to estimate the age distribution. Stari et al. (2010) extended (and included R code) the work of Gerritsen et al. (2006) by using a more flexible continuation ratio logit model. Berg and Kristensen (2012) used general additive models to develop "spatial age-length keys." Their methods are implemented in the **DATRAS** (Kristensen and Berg 2010) package.

Notes

[1]There are a number of entries in the literature regarding the bias that can result from the improper or undiscerning use of age-length keys. That literature will not be reviewed here. However, various points of entry into that literature are Kimura (1977), Westerheim and Ricker (1978), Terceiro and Ross (1993), and Goodyear (1995) for general discussions of bias and Lai (1987), Lai (1993), and Coggins Jr. et al. (2013) for discussions of appropriate and optimal sample sizes.

[2]Data manipulations in this chapter require functions from **magrittr** and **dplyr**, which are fully described in Chapter 2.

[3]Length intervals generally have a width that is a common number (e.g., 5, 10, 20, 25, or 50 mm), start at round numbers depending on the width (i.e., usually on a "0" or "5" digit), are left-inclusive and right-exclusive (i.e., the 115–120 mm interval would include 115 mm but not 120 mm fish), and are labeled with the minimum or midpoint length of the interval (e.g., the 115–120 mm interval would be labeled as either 115 or 117.5 mm). In most applications, the length intervals are of the same width, with the exception that the first and last intervals may be encompassing (e.g., "<115" or "400+"). It is not a requirement, however, that the length intervals have equal widths.

[4]Age-length keys, though slightly different than what is described here, can also be constructed with `classic_ALK()` from **ALKr** (Loff et al. 2014) and `alk()` from **fishmethods**.

[5]It may take some trial-and-error to get the value for `leg.cex=` to show the legend labels correctly.

[6]The number of fish at each age may also be obtained with matrix multiplication. Specifically, the age distribution is returned by `t(alk) %*% len.n` where `t()` transposes its argument and `%*%` is for matrix multiplication. The method with `sweep()` was used to demonstrate the steps of the computation.

[7]An approximate value for $SE(p_j)$ is provided by Lai (1987, 1993) and is commonly used in the fisheries literature. However, the only difference between it and what is provided here is that the denominator in the first fraction is n_i rather than $n_i - 1$.

[8]The `summary()` of **ALKr** and `alkprop()` of **fishmethods** perform the same calculations but use different arguments and differently formatted ALKs.

[9]Similar calculations, but from differently formatted inputs, are made with `summary()` from **ALKr**.

[10]This form of the formula is used only when the data.frame with the unaged sample already contains a variable for the assigned ages, as in this example. If the data.frame with the unaged sample did not already contain a variable for the assigned ages, then the formula should be of the form `~len`. In this case, a new variable with the assigned ages is created in the returned data.frame.

6

Size Structure

Previous chapters developed a general foundation for using R and how to analyze and prepare age data for further analysis. That foundation will be used to construct and analyze common fisheries metrics, beginning here and in Chapters 7 and 8 with analyses of fish size, before estimating population abundance in Chapters 9 and 10, and the important mortality, growth, and recruitment metrics in Chapters 11–13.

The size structure (usually, and hereafter, the distribution of lengths) of a population of fish is a synthetic response to other important ecological, population, and community factors. Neumann and Allen (2007, p. 375) described this well:

> Size structure is one of the most commonly used fisheries assessment tools. The size structure of a fish population at any point in time can be considered a snapshot that reflects the interactions of the dynamic rates of recruitment, growth, and mortality. Thus, length-frequency data provide valuable insight into the dynamics of fish populations and help identify problems such as inconsistent year-class strength, slow growth, or excessive mortality.

Furthermore, many regulations meant to protect fish populations are length-based (e.g., minimum or slot length limits; Isermann and Paukert 2010). Proper application of these regulations requires an understanding of the size structure of the population (Neumann et al. 2012). Thus, the summary metrics of size structure described in this chapter are important and common tools used by fisheries scientists.

Required Packages for This Chapter

Functions used in this chapter require loading the packages shown below.[1]

```
> library(FSA)
> library(magrittr)
> library(dplyr)
> library(plotrix)
> library(Matching)
```

6.1   Data Requirements

Size structure analyses require length measurements for individual fish. Other associated information such as species, sex, location and date of capture may also be recorded but are not required for size structure analyses. For example, *InchLake1113.csv* contains the gear type, capture year, species, and total length (in mm) for individual fish captured in 2011 to 2013 from Inch Lake (Wisconsin).

```
> inchAll <- read.csv("InchLake1113.csv")
> headtail(inchAll)
     gearType year        species  tl
1        fyke 2012 Black Crappie 246
2        fyke 2012 Black Crappie 165
3        fyke 2012 Black Crappie 330
9945     fyke 2013 Yellow Perch  97
9946     fyke 2013 Yellow Perch  84
9947     fyke 2013 Yellow Perch  79
```

The species available in this data.frame and, importantly, how they are spelled or abbreviated are found by applying `levels()` to the `species` variable.

```
> levels(inchAll$species)
[1] "Black Crappie"    "Bluegill"      "Iowa Darter"
[4] "Largemouth Bass"  "Pumpkinseed"   "Yellow Perch"
```

Primary examples in this chapter use subsets of this data.frame, specifically Bluegill captured in all years combined and separately in 2011, 2012, and 2013. These data.frames are isolated following methods from Section 2.2.2.

```
> bg <- filter(inchAll,species=="Bluegill")
> bg11 <- filter(bg,year==2011)
> bg12 <- filter(bg,year==2012)
> bg13 <- filter(bg,year==2013)
```

Several examples at the end of the chapter require data.frames of Largemouth Bass for each of 2011, 2012, and 2013. Code to construct those subsets is similar to that below but is not shown here.

Most size structure metrics require a variable that contains the length interval to which an individual belongs. These length intervals are created and appended to data.frames with `lencat()` (see Section 2.2.6) as needed throughout the chapter.

6.2 Length Frequency

A *length frequency* shows the number (i.e., frequency) of fish in all length intervals that cover the range of observed lengths. The width of the length intervals is important, as too few intervals may mask important information in the length frequency, whereas too many intervals may not provide a useful summary. Neumann et al. (2012) suggested interval widths of 10 mm for fish that reach 300 mm, 20 mm for fish that reach 600 mm, and 50 mm for fish with maximum lengths greater than 600 mm. Of course, these are suggestions, not rules, and your best judgment should be used when constructing the intervals.

The maximum length of Bluegill captured from Inch Lake in 2011 is 269 mm. Following the advice of Neumann et al. (2012), 10 mm wide length intervals were appended to the `bg11` data.frame with `lencat()`.

```
> bg11 %<>% mutate(lcat10=lencat(tl,w=10))
> headtail(bg11)
     gearType year  species  tl lcat10
1         fyke 2011 Bluegill  36     30
2         fyke 2011 Bluegill  38     30
3         fyke 2011 Bluegill  61     60
2361      fyke 2011 Bluegill 142    140
2362      fyke 2011 Bluegill 104    100
2363      fyke 2011 Bluegill  81     80
```

6.2.1 Tables

A length frequency table contains the number (i.e., frequency) of fish in each length interval. This table is constructed with `xtabs()`, with the first argument a formula of the form `~lcat`, where `lcat` is the variable that contains the length interval data, and the corresponding data.frame in `data=`.

```
> ( bgFreq10 <- xtabs(~lcat10,data=bg11) )
lcat10
  20   30   40   50   60   70   80   90  100  110  130  140  150
 237 1463  333   80   28   19   14    2    3    1    8    6    6
 160  170  180  190  200  210  220  230  240  250  260
   3    8    3    6   23   18   16   29   35   18    4
```

The percentages of fish in each length interval are computed by submitting the `xtabs()` object to `prop.table()` and multiplying the result by 100.

```
> round(prop.table(bgFreq10)*100,1)   # rounded for display only
lcat10
  20   30   40   50   60   70   80   90  100  110  130  140  150
10.0 61.9 14.1  3.4  1.2  0.8  0.6  0.1  0.1  0.0  0.3  0.3  0.3
 160  170  180  190  200  210  220  230  240  250  260
 0.1  0.3  0.1  0.3  1.0  0.8  0.7  1.2  1.5  0.8  0.2
```

Thus, 1463 Bluegill, or 61.9% of all Bluegill sampled in 2011, were 30–39 mm.

6.2.2   Histograms

A histogram shows the frequency of individuals in each length interval. Histograms are constructed with the modified `hist()` from **FSA** as described in Section 3.3. A length frequency histogram (Figure 6.1) for Bluegill captured in 2011, with custom length intervals, is constructed below.

```
> hist(~tl,data=bg11,breaks=seq(20,270,10),
        xlab="Total Length (mm)")
```

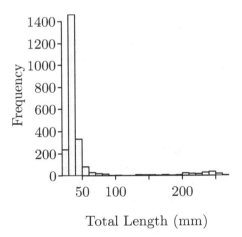

FIGURE 6.1. Length frequency histogram for Bluegill sampled from Inch Lake in 2011.

It is apparent in Figure 6.1 that a very large number of Bluegill less than 100 mm were captured. These fish can be examined more closely by creating separate histograms for those fish less than and those fish greater than 100 mm (Figure 6.2). Narrower length intervals are used for fish less than 100 mm to better highlight the shape of the distribution of the large number of fish in a relatively narrow range of lengths.

```
> hist(~tl,data=filter(bg11,tl<=100),breaks=seq(20,100,5),
      xlab="Total Length (mm)")
> hist(~tl,data=filter(bg11,tl>=100),breaks=seq(100,270,10),
      xlab="Total Length (mm)")
```

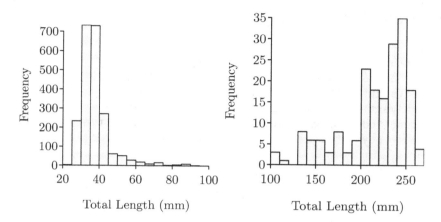

FIGURE 6.2. Length frequency histograms for Bluegill sampled from Inch Lake in 2011 that were less than (Left) and greater than (Right) 100 mm. Note that different scales and interval widths are used in the two panels.

6.2.3 Cumulative Frequencies

A length frequency may be presented as an *empirical cumulative distribution function* (ECDF; Neumann and Allen 2007), which is the proportion of fish that are less than each observed length. Plots of the ECDF, which appear "stepped" because the length data are discrete, are especially useful when comparing two or more length frequency distributions.

An ECDF is computed by giving the vector of observed lengths as the only argument to `ecdf()` and plotted by including the `ecdf()` object as the first argument to `plot()`. In this case, `plot()` takes optional arguments of `do.points=FALSE` to eliminate the plotting of points at each value (i.e., they clutter the plot), `verticals=TRUE` to vertically connect values at the discrete lengths, `main=""` to remove the unnecessary main title, and `col.01line=NULL` to remove the unnecessary horizontal lines at 0 and 1. The code below plots a single ECDF (plot not shown).

```
> plot(ecdf(bg11$tl),xlab="Total Length (mm)",do.points=FALSE,
      verticals=TRUE,main="",col.01line=NULL)
```

Plots of the ECDF are more informative when used to compare two or more samples. In this case, a plot for one sample is constructed as just described. While this plot is still active, the ECDF for another sample is added with a second call to `plot()`, but including `add=TRUE`. It is useful in the second call to change the color with `col=` to make the second ECDF distinguishable from the first ECDF. In the example below, a legend is also added to the final plot with `legend()` as described in Section 3.1.2.

```
> clr <- c("black","gray50")
> plot(ecdf(bg11$tl),xlab="Total Length (mm)",
        do.points=FALSE,verticals=TRUE,main="",col.01line=NULL)
> plot(ecdf(bg12$tl),add=TRUE,do.points=FALSE,
        verticals=TRUE,col=clr[2],col.01line=NULL)
> legend("bottomright",c("2011","2012"),col=clr,lty=1,
        bty="n",cex=0.75)
```

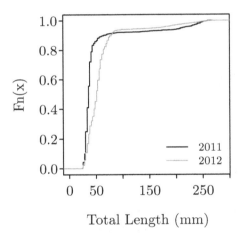

FIGURE 6.3. Empirical cumulative distribution function for Bluegill sampled in 2011 and 2012.

From Figure 6.3, it is evident that Bluegill were of approximately the same range of lengths in both years but that more Bluegill were between approximately 30 and 60 mm in 2011 than in 2012. A formal statistical comparison of the distribution of lengths between these two years is in Section 6.4.

6.3 Proportional Size Distribution (PSD)

In some instances, a fisheries scientist may desire summary metrics of size structure that are even simpler than the length frequency table and histogram. For example, interest may be in the percentage of fish that are larger than a minimum length limit value, larger than the minimum length preferred by anglers, or between two specific lengths. These percentages are simple calculations, but their interpretation is confounded by the fact that the denominator (all sampled fish) may contain small fish whose numbers may be highly variable in the first year (or so) of life, are poorly sampled by many gears, or are of little recreational or commercial value. Thus, the sample is commonly restricted to fish larger than some minimum length before calculating these percentages. That minimum length, the *stock length*, has been "variously defined as the approximate length at maturity, minimum length effectively sampled by traditional fisheries gear, and the minimum length of fish that provide recreational value" (Neumann et al. 2012, p. 642). Percentages computed from the sample of fish greater than the stock length are called *proportional size distribution (PSD) indices* (Guy et al. 2007; Neumann et al. 2012).[2]

6.3.1 Gabelhouse Length Categories

Gabelhouse (1984) defined minimum lengths for stock-, quality-, preferred-, memorable-, and trophy-sized individuals based on a percentage of world-record size for the species.[3] In his system, all fish in the "longer" categories also belong to all "shorter" categories. For example, all fish in the preferred category are also in the quality and stock categories. Gabelhouse lengths form the basis of common PSD indices introduced in the next two sections.

Gabelhouse lengths have been defined, in both English and metric units, for a large number of game fish in North America and, more recently, some nongame fish and fish from outside of North America. Gabelhouse lengths for a particular species are found by including the species name in quotes as the first argument to `psdVal()`.[4] For example, the code below shows the named vector of minimum lengths for each category for Bluegill and assigns those results to an object for later use.

```
> ( bg.cuts2 <- psdVal("Bluegill") )
 substock     stock   quality preferred memorable    trophy
        0        80       150       200       250       300
```

Thus, all Bluegill longer than 80 mm are considered to be of stock-length and those longer than 200 mm are also considered to be of preferred-length.

Gabelhouse lengths may be retrieved in alternative formats from `psdVal()`. For example, `units=` controls whether the results are returned in millimeters (`="mm"`; default), centimeters (`="cm"`), or inches (`="in"`) and incl.zero= controls whether the list of values returned will include a zero as the first value in the vector (=TRUE; default) or not (=FALSE). Including a zero in the vector is particularly useful when fish smaller than the stock length are present in the data and are also of interest to the scientist (this is why the zero value is named "substock").

Gabelhouse length categories are appended to a data.frame by including the psdVal() object in breaks= of lencat(). Because the psdVal() object is a named vector, using use.names=TRUE will label the length categories with the category name rather than length (see Section 2.2.6). Length categories that contain no fish are excluded from consideration with drop.levels=TRUE (e.g., if no trophy-length fish are in the data, then the trophy category will be excluded). The code below creates a new data.frame with just stock-length (and longer) Bluegill and appends the Gabelhouse length categories in the new variable gcat.

```
> bg11s <- bg11 %>%
    filter(tl>=bg.cuts2["stock"]) %>%
    mutate(gcat=lencat(tl,breaks=bg.cuts2,
                    use.names=TRUE,drop.levels=TRUE))
> headtail(bg11s)
    gearType year  species  tl lcat10  gcat
1       fyke 2011 Bluegill 104    100 stock
2       fyke 2011 Bluegill  81     80 stock
3       fyke 2011 Bluegill 130    130 stock
201     fyke 2011 Bluegill 142    140 stock
202     fyke 2011 Bluegill 104    100 stock
203     fyke 2011 Bluegill  81     80 stock
```

Thus, for example, the frequency of fish in each Gabelhouse length interval is computed from this new variable with xtabs().

```
> ( gfreq <- xtabs(~gcat,data=bg11s) )
gcat
    stock   quality preferred memorable
       34        26       121        22
```

The frequencies in each length interval are converted to percentages with prop.table().

```
> ( psdXY1 <- prop.table(gfreq)*100 )
gcat
    stock   quality preferred memorable
 16.74877  12.80788  59.60591  10.83744
```

Thus, 34 Bluegill, or 16.7% of all *stock-length* Bluegill, are greater than or equal to the stock (80 mm) and less than the quality (150 mm) lengths.

6.3.2 Incremental PSD Indices

An *incremental PSD* is the percentage of all stock-length fish that are between two successive Gabelhouse lengths (Neumann et al. 2012). For example, $PSD\ Q-P$ is the percentage of all stock-length fish that are greater than or equal to the quality length and less than the preferred length. Similarly, $PSD\ S-Q$ is the percentage of all stock-length fish that are greater than or equal to the stock length and less than the quality length. Thus, generically,

$$PSD\ X-Y = \frac{\text{Number of fish} \geq \text{length } X \text{ but} < \text{length Y}}{\text{Number of fish} \geq \text{stock length}} * 100$$

These values were computed with `xtabs()` and `prop.table()` at the end of the previous section. Thus, for example, $PSD\ S-Q$ is 17. Similarly, $PSD\ Q-P$ is 13, $PSD\ P-M$ is 60, and $PSD\ M-T$ is 11. As seen here, PSD indices are usually rounded to the nearest integer and presented without the percentage symbol (Neumann et al. 2012).

6.3.3 Traditional PSD Indices

The PSD indices were originally defined (Anderson 1976) as

$$PSD-X = \frac{\text{Number of fish} \geq \text{length } X}{\text{Number of fish} \geq \text{stock length}} * 100$$

where X is replaced with a descriptor specific to the length in the numerator. For example, $PSD-Q$ uses the number of fish greater than or equal to the quality length as the numerator and would be interpreted as the percentage of stock-length fish that are also quality length. Similarly, $PSD-P$ uses the preferred length in the numerator and is the percentage of stock-length fish that are also preferred length. $PSD-Q$ is the most commonly calculated value and is often referred to as simply PSD (Guy et al. 2007).[5]

The $PSD-X$ value is also equal to the sum of all $PSD\ X-Y$ values for lengths greater than X. For example, $PSD-Q$ is the sum of $PSD\ Q-P$, $PSD\ P-M$, $PSD\ M-T$, and $PSD-T$. Thus, the $PSD-X$ values are computed as the *reverse* cumulative sum (i.e., the sum "from the right" rather than "from the left") of the percentage of fish in each Gabelhouse length interval. The reverse cumulative sum of a vector is computed with `rcumsum()` from **FSA**.

```
> ( psdX1 <- rcumsum(psdXY1) )
    stock   quality preferred memorable
100.00000  83.25123  70.44335  10.83744
```

Thus, $PSD-Q$, the percentage of stock-length fish that are also quality length, is 83.

6.3.4 Special PSD Indices

There are instances when the fisheries scientist is interested in lengths other than the Gabelhouse lengths. For example, one may want to compute the percentage of stock-length fish that are above a minimum length limit or within a (protected or harvest) slot limit. These "special" PSD indices are computed by either manually defining a vector of minimum category lengths or, if interest also includes some of the Gabelhouse length categories, using the addLens= argument in psdVal(). A custom vector that contains the stock-length and special interest values of 175 and 225 mm is shown below (the use of the addLens= argument is illustrated with a related function in Section 6.3.6).

```
> bg11s %<>% mutate(gcat2=lencat(tl,breaks=c(80,175,225)))
> gfreq2 <- xtabs(~gcat2,data=bg11s)
> ( psdXY2 <- prop.table(gfreq2)*100 )
gcat2
       80      175      225
23.64532 29.06404 47.29064
> ( psdX2 <- rcumsum(psdXY2))
        80      175      225
100.00000  76.35468  47.29064
```

Thus, for example, if this population of Bluegill was being managed with a protected slot limit from 175 to 225 mm, then 29% of the sampled fish were within the slot limit ($PSD\ 175-225$) and 47% were above the slot limit ($PSD-225$).

6.3.5 Confidence Intervals

Confidence intervals for PSD indices may be computed in two ways. First, if interest is in more than one index value (e.g., $PSD\ S-Q$, $PSD\ Q-P$, and $PSD-Q$), which is most often the case, then the multinomial distribution is used (Brenden et al. 2008).[6] If interest lies in only one PSD index value, then the confidence interval is computed with the binomial distribution (Gustafson 1988).[7]

Brenden et al. (2008) outlined a flexible procedure to compute confidence intervals for PSD indices. This procedure is applied in a fairly involved manner in the remainder of this section. A function that simplifies the implementation of this procedure, however, is introduced in Section 6.3.6. Some readers may skip this section if they only want to use the more convenient function.

The flexibility of the procedure outlined in Brenden et al. (2008) is afforded through a vector of zeroes and ones that identify the $PSD\ X-Y$ values that will form the PSD index value of interest. A 1 indicates the position in the summary percentages table (i.e., the position of the $PSD\ X-Y$ value) to be used, whereas a 0 indicates a position not used. For example, with the four Gabelhouse lengths that exist in the 2011 Bluegill results, a vector of c(1,0,0,0) would indicate that only the first value is used (i.e., PSD $S-Q$). Similarly, a vector of c(0,0,1,0) would indicate that only the third value is used (i.e., PSD $P-M$). However, c(0,0,1,1) indicates that the third **AND** fourth values (PSD $P-M$ and PSD $M-T$) are used, which are summed to form $PSD-P$. One must be careful to ensure that the vector contains only as many zeroes and ones as there are Gabelhouse lengths for which fish exist in the sample (e.g., four in this case).

This procedure and use of the binomial and multinomial distributions are implemented in psdCI() from **FSA**, where the vector of indicator values is the first argument, the table with the percentages of fish in each length interval is given in ptbl=, and the total number of sampled stock-length fish is given in n=. The distribution to use is selected with method="binomial" or method="multinomial". For convenience, the row of the result may be labeled with a string given in label=.

The code below computes a confidence interval for only PSD $S-Q$.

```
> psdCI(c(1,0,0,0),ptbl=psdXY1,n=sum(gfreq),method="binomial",
         label="PSD S-Q")
         Estimate 95% LCI 95% UCI
PSD S-Q      16.7    12.2    22.5
```

Thus, between 12% and 22% of all stock-length Bluegill are greater than or equal to the stock (80 mm) and less than the quality (150 mm) lengths.

Similarly, the code below computes a confidence interval for $PSD-Q$.

```
> psdCI(c(0,1,1,1),ptbl=psdXY1,n=sum(gfreq),method="binomial",
         label="PSD-Q")
         Estimate 95% LCI 95% UCI
PSD-Q        83.3    77.5    87.8
```

Thus, between 78% and 88% of all stock-length Bluegill are greater than or equal to the quality (150 mm) length.

Confidence intervals for multiple PSD indices may be computed by "row-binding" (with rbind()) vectors of indicator values into a matrix and then using apply() to cycle through those rows. For example, the matrix of indicator values required to compute simultaneous confidence intervals for PSD $S-Q$ and PSD $Q-P$ in the Bluegill example is constructed below (including how to label the rows).[8]

```
> ( ivmat <- rbind("PSD S-Q"=c(1,0,0,0),
                    "PSD Q-P"=c(0,1,0,0)) )
          [,1] [,2] [,3] [,4]
PSD S-Q     1    0    0    0
PSD Q-P     0    1    0    0
```

The rows of this matrix are sequentially submitted to `psdCI()` with `apply()` as described in Section 1.9.2. Thus, the code below submits the rows of the `ivmat` matrix to `psdCI()`. More than one confidence interval is constructed here so the multinomial distribution must be declared with `method="multinomial"`. The matrix returned by `apply()` is oriented opposite of what one generally wants, so it is transposed with `t()`. Finally, the columns of the resulting matrix are named with `colnames()`.

```
> psdXY2 <- t(apply(ivmat,FUN=psdCI,MARGIN=1,
                    ptbl=psdXY1,n=sum(gfreq),
                    method="multinomial"))
> colnames(psdXY2) <- c("Estimate","95% LCI","95% UCI")
> psdXY2
          Estimate 95% LCI 95% UCI
PSD S-Q       16.7     9.4    24.1
PSD Q-P       12.8     6.3    19.4
```

6.3.6   Easy Calculations of Indices and Confidence Intervals

The process for constructing simultaneous confidence intervals for PSD indices illustrated in Section 6.3.5 is more conveniently completed for **all** PSD indices with `psdCalc()` from **FSA**. This function takes a formula of the form `~length` as the first argument, the corresponding data.frame in `data=`, the name of the species in `species=` (as in `psdVal()`), and the units of measurement in `units=` (as in `psdVal()`). The incremental PSD values are returned by including `what="incremental"`, whereas traditional PSD values are returned by including `what="traditional"`.

```
> psdCalc(~tl,data=bg11,species="Bluegill",what="incremental")
          Estimate 95% LCI 95% UCI
PSD S-Q         17       9      24
PSD Q-P         13       6      19
PSD P-M         60      50      69
PSD M-T         11       5      17

> psdCalc(~tl,data=bg11,species="Bluegill",what="traditional")
          Estimate 95% LCI 95% UCI
PSD-Q           83      76      91
PSD-P           70      61      79
PSD-M           11       5      17
```

In instances when the fisheries scientist is interested in lengths other than the Gabelhouse lengths, these additional lengths may be included in a vector and submitted to `addLens=` in `psdCalc()`. Names for the additional lengths will be used if a named vector is given to `addLens=` as illustrated below. If names are in a separate vector, then this vector is given to `addNames=` (demonstrated in the `psdCalc()` help documentation).[9] The code below demonstrates the addition of two length categories and that `psdCalc()` will return all PSD indices if `what=` is not given.

```
> psdCalc(~tl,data=bgl1,species="Bluegill",
          addLens=c(minSlot=175,maxSlot=225))
```

```
Warning: Some category sample size <20, some CI coverage may be
lower than 95%.
              Estimate 95% LCI 95% UCI
PSD-Q               83      75      92
PSD-minSlot         76      66      86
PSD-P               70      60      81
PSD-maxSlot         47      36      59
PSD-M               11       4      18
PSD S-Q             17       8      25
PSD Q-minSlot        7       1      13
PSD minSlot-P        6       0      11
PSD P-maxSlot       23      13      33
PSD maxSlot-M       36      25      48
PSD M-T             11       4      18
```

Following the advice of Brenden et al. (2008), a warning may be issued if the sample size in any category falls below 20 individuals. Also note that `psdCalc()` is especially convenient because it works on the "raw" length data and does not require any preprocessing (e.g., filtering to stock-length fish, mutating a length category variable, or creating a table of percentages).

6.3.7   Tic-Tac-Toe Plots

A *tic-tac-toe plot* is used to illustrate paired PSD values, usually from predator-prey pairs, relative to target objectives. Target values are usually defined based on management or ecological objectives (Neumann et al. 2012). For example, Willis et al. (1993) suggested that the $PSD - Q$ for Largemouth Bass should be 50–80 and the $PSD - Q$ for Bluegill should be 10–50 if the management objective (for small ponds) was for "big bass."

Before constructing the tic-tac-toe plot, the PSD values and associated confidence intervals must be computed and assigned to separate matrices for the two groups. In the example below, `psdCalc()` is applied separately to each year of Bluegill data.

```
> psd.bgl1 <- psdCalc(~tl,data=bgl1,species="Bluegill")
> psd.bgl2 <- psdCalc(~tl,data=bgl2,species="Bluegill")
> psd.bgl3 <- psdCalc(~tl,data=bgl3,species="Bluegill")
```

The $PSD - Q$ values are extracted from the PSD-Q row of each of these results and row-bound together with rbind() to form the required matrix for Bluegill.

```
> ( psd.bg <- rbind(psd.bg11["PSD-Q",],psd.bg12["PSD-Q",],
                    psd.bg13["PSD-Q",]) )
     Estimate 95% LCI 95% UCI
[1,]       83      76      91
[2,]       59      51      67
[3,]       24      19      29
```

The process is repeated for Largemouth Bass (construction of the data.frames for Largemouth Bass was not shown).

```
> psd.lmb11 <- psdCalc(~tl,data=lmb11,species="Largemouth Bass")
> psd.lmb12 <- psdCalc(~tl,data=lmb12,species="Largemouth Bass")
> psd.lmb13 <- psdCalc(~tl,data=lmb13,species="Largemouth Bass")
> ( psd.lmb <- rbind(psd.lmb11["PSD-Q",],psd.lmb12["PSD-Q",],
                     psd.lmb13["PSD-Q",]) )
     Estimate 95% LCI 95% UCI
[1,]       86      60     100
[2,]       74      59      90
[3,]       29      11      47
```

A base tic-tac-toe plot is constructed with tictactoe() from **FSA**. By default, the target objective for both the predator and prey is 30–70. The target objectives are changed by including vectors with the objective values in predobj= and preyobj=, respectively. The example below creates a base tic-tac-toe plot with target objectives for the "big bass" scenario defined by Willis et al. (1993).

```
> tictactoe(predobj=c(50,80),predlab="Largemouth Bass PSD-Q",
            preyobj=c(10,50),preylab="Bluegill PSD-Q")
```

Points for the paired PSD values, and their associated confidence intervals, are added to the base plot with plotCI() from **plotrix**.[10] The vectors for the x- and y-coordinates of the points are the first two arguments to plotCI(). The lower and upper bounds for the confidence intervals are given to li= and ui=, respectively, and the direction of the intervals is controlled with err= (set to either "x" or "y"). Finally, add=TRUE is used so that the points and intervals are added to the active plot (initially, the base tic-tac-toe plot). The example below also uses pch=19 to set the plotting character to a filled but outlined circle. Note that plotCI() must be called twice in this example, once to add the intervals for the Largemouth Bass (in the x direction) and once for Bluegill (in the y direction).

```
> # add predator (lmb) error in x direction
> plotCI(psd.lmb[,"Estimate"],psd.bg[,"Estimate"],
        li=psd.lmb[,"95% LCI"],ui=psd.lmb[,"95% UCI"],
        err="x",add=TRUE,pch=19)
> # add prey (bg) error in y direction
> plotCI(psd.lmb[,"Estimate"],psd.bg[,"Estimate"],
        li=psd.bg[,"95% LCI"],ui=psd.bg[,"95% UCI"],
        err="y",pch=19,add=TRUE)
```

The points in the plot may be connected with lines() (see Section 3.2.2) to show the trajectory of values over time. Labels may be placed next to the points with text() (see Section 3.6.2). By default, the labels are centered on the coordinates (in this case, over the points). However, the two values in adj= are used to adjust the x and y positioning of the points, respectively (this usually requires trial-and-error to position the points where you desire).[11] A cex= value less than 1 reduces the size of the text.

```
> lines(psd.lmb[,"Estimate"],psd.bg[,"Estimate"])
> text(psd.lmb[,"Estimate"],psd.bg[,"Estimate"],
        labels=2011:2013,adj=c(1.2,-0.5),cex=0.75)
```

The final result is the tic-tac-toe plot in Figure 6.4. The results show that both species were initially above their target objectives for $PSD - Q$ (i.e., a higher percentage of "large" fish) and Bluegill had reached their target objective by 2013 but Largemouth Bass were below their target objective (i.e., a higher percentage of "smaller" fish) at that time.

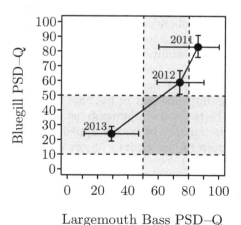

FIGURE 6.4. Tic-tac-toe plot for the $PSD - Q$ of Largemouth Bass and Bluegill sampled from Inch Lake in 2011–2013.

6.4 Among Group Statistical Comparisons

6.4.1 Kolmogorov-Smirnov Test

Length frequency distributions can be compared with the Kolmogorov-Smirnov (K-S) two-sample test (Neumann and Allen 2007). The K-S test is used to determine whether the ECDF (Section 6.2.3) are the same between two groups and can detect differences in the location (e.g., median), dispersion (e.g., variance), and shape of the distributions (Hollander et al. 2014).

The K-S test is performed with ks.test() which requires vectors of lengths from the two samples as the first two arguments.

```
> ks.test(bg11$tl,bg12$tl)

Warning in ks.test(bg11$tl, bg12$tl): p-value will be
approximate in the presence of ties

Two-sample Kolmogorov-Smirnov test

data:  bg11$tl and bg12$tl
D = 0.48374, p-value < 2.2e-16
alternative hypothesis: two-sided
```

This result suggests that there is a strong difference ($p < 0.00005$) in length distributions between Bluegill captured in 2011 and those captured in 2012.

One should, however, be cognizant of the warning given by ks.test(). The K-S test was constructed to work with *continuous* data where ties are theoretically impossible (Hollander et al. 2014). This assumption is often violated with discrete length data and, thus, one should interpret the results of the K-S test cautiously, especially for samples with narrow ranges of observed lengths.

A bootstrapped version of the K-S test that is insensitive to ties with noncontinuous data is implemented in ks.boot() from **Matching** (Sekhon 2011).[12] In addition to the same arguments as ks.test(), ks.boot() uses nboots= to control the number of bootstrapped samples. The bootstrapped p-value is extracted from the ks.boot() object with summary().

```
> bg.ksb <- ks.boot(bg11$tl,bg12$tl,nboots=5000)
> summary(bg.ksb)

Bootstrap p-value:     < 2.22e-16
Naive p-value:         0
Full Sample Statistic: 0.48374
```

The bootstrapped p-value also suggests strong evidence for a difference in length distributions between Bluegill captured in the two years ($p < 0.00005$).

If the fisheries scientist is comparing more than two years, then each pair of years must be compared separately and the p-values from each comparison must be adjusted for an increasing experimentwise error rate due to multiple comparisons. P-values are adjusted for multiple comparisons with p.adjust(), which takes a vector of unadjusted p-values as its first argument.[13] The p-value for a single K-S test is extracted by appending $p.value to a ks.test() object or $ks.boot.pvalue to a ks.boot() object. As an example, p-values are extracted from ks.test() object, combined into a vector, and adjusted with p.adjust() below.

```
> ( ks.ps <- c(ks.test(bg11$tl,bg12$tl)$p.value,
               ks.test(bg11$tl,bg13$tl)$p.value,
               ks.test(bg12$tl,bg13$tl)$p.value) )
[1] 0 0 0
> p.adjust(ks.ps)
[1] 0 0 0
```

The adjusted p-values from all pairwise comparisons are very small (less than the smallest number that R will report) which indicates that the length distribution of Bluegill differs for each year from 2011–2013.

6.4.2 Chi-Square Test

A chi-square test may be used to test for differences in length frequencies among different groups (e.g., among times, locations, or gears; Neumann and Allen 2007). Most often, the length intervals used for a chi-square test are broad such that all or most cells in the table are expected to have more than five individuals.[14]

As an example, the frequency of stock-length Bluegill captured in 2011 through 2013 are examined. A data.frame restricted to stock-length fish and with the Gabelhouse length intervals is constructed first.

```
> # note: bg.cuts2 was constructed previously
> bgs <- bg %>% filter(tl>=bg.cuts2["stock"]) %>%
      mutate(gcat=lencat(tl,breaks=bg.cuts2,
                         use.names=TRUE,drop.levels=TRUE))
```

A two-way table of frequencies in each length interval for each group must be constructed to perform the chi-square test in R. These two-way tables are constructed with xtabs() using a formula of the form ~row+col, where row and col generically represent the factor variables that will form the rows and columns, respectively, of the resulting table. A table of frequencies of stock-length Bluegill in the Gabelhouse lengths for all three years is constructed below.

```
> ( bg.LF <- xtabs(~year+gcat,data=bgs) )
      gcat
year    stock quality preferred memorable
  2011    34      26       121        22
  2012   114      69        75        19
  2013   423     104        27         2
```

The chi-square test is conducted by submitting this table to chisq.test().

```
> chisq.test(bg.LF)

        Pearson's Chi-squared test

data:  bg.LF
X-squared = 375.95, df = 6, p-value < 2.2e-16
```

The small p-value ($p < 0.00005$) indicates that the distribution of fish into the length intervals differs among the three years. A table of row percentages (i.e., the $PSD\ X-Y$ values for each year) can help identify where differences may exist. A row percentage table is computed by submitting the frequency table to prop.table() as done previously, but including margin=1 to indicate that the proportions are computed by dividing by the total of each row.

```
> round(prop.table(bg.LF,margin=1)*100,0)
      gcat
year    stock quality preferred memorable
  2011    17      13        60        11
  2012    41      25        27         7
  2013    76      19         5         0
```

From this, it appears that the percentages of preferred- and memorable-length fish declined and the percentage of stock-length fish increased from 2011 to 2013.

While it did not happen here, the chi-square test will fail in most instances where the contingency table contains zeroes or very small frequencies in several cells.[15] One way to handle this issue is to combine adjacent length categories. For example, one may combine the memorable-length fish with the preferred-length fish to form a "preferred or larger" interval. One way to combine these length intervals is to use mapvalues() as described in Section 2.2.7. When mapvalues() is used to reduce the number of levels, it should be followed immediately by droplevels() to make sure that R adjusts the possible levels.

```
> bgs %<>% mutate(gcat2=mapvalues(gcat,
                    from=c("preferred","memorable"),
                    to=c("preferred+","preferred+")),
                  gcat2=droplevels(gcat2))
> ( bg.LF2 <- xtabs(~year+gcat2,data=bgs) )
      gcat2
year   stock quality preferred+
  2011    34      26        143
  2012   114      69         94
  2013   423     104         29
```

Finally, a fisheries scientist may want to test for a difference in $PSD-X$ values between groups. For example, one may want to determine if $PSD-Q$ values differed among years. In these cases, the frequency table must be constructed from a variable that contains only two levels, those at or above X and those below X. This recategorization is again accomplished with mapvalues() and droplevels(). For example, the chi-square test below is used to determine if $PSD-Q$ values for Bluegill differed significantly among sample years.

```
> bgs %<>% mutate(gcat3=mapvalues(gcat,
            from=c("stock","quality","preferred","memorable"),
            to=c("quality-","quality+","quality+","quality+")),
          gcat3=droplevels(gcat3))
> ( bg.LF3 <- xtabs(~year+gcat3,data=bgs) )
      gcat3
year   quality- quality+
  2011       34      169
  2012      114      163
  2013      423      133
> chisq.test(bg.LF3)

 Pearson's Chi-squared test

data:  bg.LF3
X-squared = 241.39, df = 2, p-value < 2.2e-16
```

The small p-value ($p < 0.00005$) indicates that the $PSD-Q$ of Bluegill for at least one year is different from the other years.

Multiple comparisons are conducted by isolating each pair of years, accumulating the p-values from the chi-square test for each pair of years (p-values are extracted by appending $p.value to the chisq.test() object), and then adjusting the p-values with p.adjust(). Note below that the first argument to chisq.test() is pairs of rows (i.e., years) selected from the frequency table that had all three years.

```
> ( ps <- c(chisq.test(bg.LF3[1:2,])$p.value,
             chisq.test(bg.LF3[c(1,3),])$p.value,
             chisq.test(bg.LF3[2:3,])$p.value) )
[1] 1.907395e-08 6.654710e-49 7.207540e-23
> p.adjust(ps)
[1] 1.907395e-08 1.996413e-48 1.441508e-22
```

From this, it appears that there is strong evidence for a difference in $PSD-Q$ values between all years (i.e., all p-values are very small).

6.5 Further Considerations

Conveniently Summarizing PSD for ALL Species

The process for computing PSDs described in Section 6.3 requires filtering a data.frame to isolate only one species. This is cumbersome if one has several species to summarize. A variable with Gabelhouse length intervals for **all** species (for which Gabelhouse length definitions exist) in a data.frame is created with psdAdd(). When used with mutate(), the first two arguments to psdAdd() are variables that contain the length measurements and the name of the species. Important optional arguments are units= and use.names= (TRUE is the default here) as described for psdVal(). Finally, additional lengths (beyond the Gabelhouse lengths) for a species are included by including the lengths in a vector given to addLens= and the species names (in quotes) in a corresponding vector given to addSpec=. For example, simply adding the Gabelhouse lengths to the original inchAll data.frame is illustrated below.

```
> inchAll %<>% mutate(gcat=psdAdd(tl,species))
```
No known Gabelhouse (PSD) lengths for Iowa Darter
```
> headtail(inchAll)
     gearType year       species  tl      gcat
1        fyke 2012 Black Crappie 246   quality
2        fyke 2012 Black Crappie 165     stock
3        fyke 2012 Black Crappie 330  memorable
9945     fyke 2013  Yellow Perch  97  substock
9946     fyke 2013  Yellow Perch  84  substock
9947     fyke 2013  Yellow Perch  79  substock
```

Individual species still need to be isolated to construct confidence intervals and perform hypothesis tests as illustrated above. However, summaries of PSD values for all species may be efficiently computed from the data.frame with all species. All substock fish should be removed, though, before creating this summary.

```
> inchAlls <- filterD(inchAll,gcat!="substock")
```

Incremental PSD values for each species are quickly computed with xtabs() and prop.table().

```
> freq <- xtabs(~species+gcat,data=inchAlls)
> iPSDs <- prop.table(freq,margin=1)*100
> round(iPSDs,0)
                  gcat
species        stock quality preferred memorable
  Black Crappie    20      25         5        50
  Bluegill         55      19        22         4
  Largemouth Bass  42      58         0         0
  Pumpkinseed      71      29         0         0
  Yellow Perch     80       0         0        20
```

Traditional PSD values for each species are computed by sending each row of the table of incremental PSD values to rcumsum() with apply() (see Section 1.9.2 for a description of apply()).

```
> PSDs <- t(apply(iPSDs,MARGIN=1,FUN=rcumsum))
> round(PSDs,1)
                  gcat
species        stock quality preferred memorable
  Black Crappie   100    79.6      54.9      50.0
  Bluegill        100    44.9      25.7       4.2
  Largemouth Bass 100    58.1       0.0       0.0
  Pumpkinseed     100    28.6       0.0       0.0
  Yellow Perch    100    20.0      20.0      20.0
```

It is also possible to summarize these data by species **AND** year; though, this is left largely as an exercise for the reader.

```
> freq <- xtabs(~year+species+gcat,data=inchAlls)
> ffreq <- ftable(freq)
> iPSDs <- prop.table(ffreq,margin=1)*100
> round(iPSDs,1)
> PSDs <- t(apply(iPSDs,MARGIN=1,rcumsum))
> round(PSDs,1)
```

Notes

[1]Data manipulations in this chapter require functions from **magrittr** and **dplyr**, which are fully described in Chapter 2.

[2]The name of these indices is confused in the literature. The indices described here were originally called the proportional stock density (also abbreviated as PSD) and relative stock density (RSD). Guy et al. (2006) changed these two names to proportional size structure (PSS). However, Guy et al. (2007) then changed these names to proportional size distribution, which is what is used here. It appears that the use of "proportional size distribution" has not taken hold, as many authors still use "proportional stock density" and the same acronym further confuses the situation. Neumann et al. (2012) provided a nice description with a tabular comparison of acronyms that should help clarify this confusion. The PSD acronym represents "proportional size distribution" throughout this book.

[3]For example, Gabelhouse (1984) defined quality-length fish as those fish that are at least 36% of the world-record length for that species and that most anglers would like to catch.

[4]All species with Gabelhouse lengths in `psdVal()` may be seen by typing `psdVal()` (i.e., without any arguments). The entire data.frame of information regarding the Gabelhouse lengths is seen by first typing `data(PSDlit)` and then `PSDlit`.

[5]However, this convention will not be used in this book. Thus, the "PSD" acronym will be used when discussing the general method, but $PSD-Q$ will be used for the percentage of stock-length fish that are also quality length.

[6]The multinomial distribution is an extension of the binomial distribution where more than two choices are possible. The implementation of the multinomial distribution is used when more than one PSD index value is of interest because the resulting confidence intervals provide simultaneous coverage at the desired confidence level, rather than a lower coverage that results from compounding errors from multiple uses of the binomial distribution (Brenden et al. 2008). Thus, the use of the multinomial distribution is controlling the experimentwise error rate, similar to the methods described in Sections 7.5 and 8.3.2.

[7]The binomial distribution theory applies here because each stock-length fish is recorded as having been of the specific length in the numerator or not (i.e., two choices) and the proportion of "successes" (i.e., of the specific length) is of interest to the fisheries scientist. Gustafson (1988) described the normal approximation to the binomial distribution and is often cited in the fisheries literature. With modern computing methods, there is no reason to continue to use those approximations.

[8]Note that the quotes are needed here because of the spaces and the hyphen in the row labels.

[9]If no names are given with either option, then the length value will be used as a name.

[10]Several packages have functions called `plotCI()`. If those other packages are loaded after **plotrix**, then simply typing `plotCI()` may attempt to use the "wrong" `plotCI()`. The **plotrix** version will be used with `plotrix::plotCI()`.

[11]The two values in the vector given to `adj=` contain values that identify the justification in the x and y directions, respectively, relative to the point. In the x direction, 0 means "left-justified" which means that the leftmost portion of the text will appear at the supplied x-coordinate. Furthermore, in the x direction, 1 means "right-justified" and 0.5 means centered. In the y direction, 0 means "bottom-justified", which means that the bottom edge of the text will appear at the supplied y-coordinate. Furthermore, in the y direction, 1 means "top-justified" and 0.5 means centered. Thus, `adj=c(0,0)` would place the text to the right and above the point with the bottom-left edge of the text right on the point. Using `adj=c(0.5,0.5)` (the default) would center the text right on the point. To make things even more interesting, values outside of [0,1] can also be used. One should practice with these values, but, for example, `adj=c(-0.5,-0.5)` gives an "extreme" left- and bottom-justification such that the label is to the right and above the point, but not touching the point.

[12]In `ks.boot()`, all data from both groups are combined to form one group, two groups of

data are then randomly constructed from the combined data, the K-S test is applied to the two new groups of data, the test statistic from this application is extracted, and this process is repeated a large number of times (1000, by default). The bootstrapped p-value is then the proportion of test statistics from the bootstrapped samples that exceed the test statistic computed from the original observed data.

[13]There are many algorithms to adjust p-values for multiple comparisons. A discussion of these algorithms is beyond the scope of this book. However, several of these algorithms are implemented in p.adjust() and may be selected for use with method=. See the help documentation for p.adjust() for more details. However, note that the "holm", "hochberg", and "hommel" methods are all sequential Bonferroni methods that are superior to the familiar "bonferroni" method in terms of controlling the experimentwise error rate. The "BH" and "BY" methods control the "false discovery rate" rather than the experimentwise Type I error rate.

[14]References disagree on the assumptions for expected values in a chi-square test. However, one commonly used assumption is that all expected counts exceed five (Sokal and Rohlf 1995); however, see Fienberg (1980) and Yates et al. (1999). The expected counts for each cell are found by attaching $expected to the chisq.test() object.

[15]As mentioned previously, the actual assumption of the chi-square test is that there are more than five individuals in each cell of the *expected* table. However, fewer than five individuals in the expected table will often occur when the observed table is sparse. A warning is issued by chisq.test() if any cell in the expected table contains fewer than five individuals.

7

Weight-Length Relationships

Modeling the relationship between fish weight and length has been scorned as having little value (Hilborn and Walters 2001) or been considered a routine analysis for which the results do not warrant publication (Froese 2006). However, the review and meta-analysis by Froese (2006) demonstrated that a synthetic analysis of weight-length relationships can provide important ecological insights. Jellyman et al. (2013, p. 450) summarized this importance:

> Understanding how fish weight changes as a function of length is fundamental information for fisheries scientists trying to deduce age structure, calculate growth rates, model bioenergetics or quantify some other aspect of fish population dynamics. [Weight-length relationships] are required for: 1) estimating weight from fish lengths when time or technical constraints means they cannot be recorded in the field; 2) use in stock assessment models when converting growth in length to growth in weight; 3) estimating biomass of a fish community using only length and species data; 4) estimating fish condition factor; and 5) making comparisons of fish life history characteristics.

Weight-length relationships are most often modeled with simple linear or dummy variable regressions. Thus, this chapter also serves as an introduction to these statistical tools. Note, however, that the focus here is on how to fit, assess, and interpret linear regression models in R, whereas the statistical theory underlying them is only briefly discussed.[1]

Required Packages for This Chapter

Functions used in this chapter require loading the packages shown below.[2]

```
> library(FSA)
> library(car)        # Before dplyr to reduce conflicts with MASS
> library(magrittr)
> library(dplyr)
```

7.1   Data Requirements

The data required to examine fish weight-length relationships are measurements of the length (L) and weight (W) of individual fish. Length measurements may be total, fork, or standard lengths and weights may be wet or dry, whole or dressed (Jennings et al. 2012). Length measurements are generally very precise, whereas weight measurements, especially those taken in the field, can vary substantially (Gutreuter and Krzoska 1994). Thus, some small fish may be weighed too imprecisely for use in weight-length analyses (Neumann et al. 2012). Other data about individual fish (e.g., date or location of capture) may also be recorded, but those data are not required for assessing the weight-length relationship. Generally, fish from dates and areas in close proximity should be analyzed together to reduce variability.

Weights (g) and total lengths (mm) from Ruffe captured in the St. Louis River Harbor (Minnesota and Wisconsin) are used throughout this chapter. These data are from Ogle and Winfield (2009) and are in *RuffeSLRH.csv*. To eliminate within-season variability, only Ruffe captured in July are used in this chapter. Additionally, for purposes that are explained later, the common logarithms of weight and length (computed with `log10()`) are appended to the data.frame. Finally the `fishID` and `day` variables were removed to save space in the output.

```
> ruf <- read.csv("RuffeSLRH.csv") %>%
    filter(month==7) %>%
    mutate(logW=log10(wt),logL=log10(tl)) %>%
    select(-fishID,-day)
> headtail(ruf)
     year month  tl   wt       logW      logL
1    1988     7  78  6.0 0.7781513 1.892095
2    1988     7  81  7.0 0.8450980 1.908485
3    1988     7  82  7.0 0.8450980 1.913814
1936 2004     7 137 28.0 1.4471580 2.136721
1937 2004     7 143 31.4 1.4969296 2.155336
1938 2004     7 174 82.4 1.9159272 2.240549
```

Finally, only Ruffe captured in 1990 are used in Section 7.3, whereas Ruffe captured in 1990 and 2000 are used in Section 7.4.

```
> ruf90 <- filterD(ruf,year==1990)
> ruf9000 <- filterD(ruf,year %in% c(1990,2000))
```

7.2 Weight-Length Model

There are two reasons why it is inappropriate to use linear regression to model the *raw* weight-length data for most species. First, the relationship between fish weight and length is often nonlinear (e.g., Figure 7.1-Left) because most fish add a linear amount of length but a three-dimensional volume of mass. Second, the weights of shorter fish are often less variable than the weights of larger fish (e.g., the vertical scatter of points increases from left to right in Figure 7.1-Left).

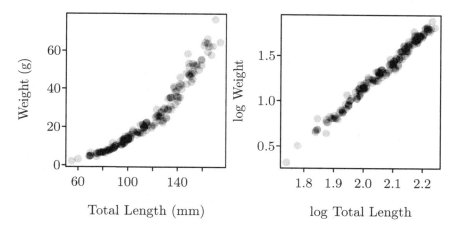

FIGURE 7.1. Untransformed (Left) and log_{10}-log_{10} transformed (Right) weight and length of Ruffe captured in 1990.

These two characteristics suggest that the weight-length relationship for most species will follow a two-parameter power function with a multiplicative error term.[3] Specifically,

$$W_i = \alpha L_i^{\beta} 10^{\epsilon_i} \tag{7.1}$$

where α and β are parameters and 10^{ϵ_i} is the multiplicative error term for the ith fish. Equation (7.1) is *transformed* to a linear model by applying common logarithms[4] to both sides and simplifying

$$log_{10}(W_i) = log_{10}(\alpha) + \beta log_{10}(L_i) + \epsilon_i \tag{7.2}$$

In addition to being linear, Equation (7.2) has errors that are additive and will show a constant variability around the line for all lengths (Figure 7.1-Right). Thus, linear regression methods are used to fit a linear model to log-transformed weight-length data, where the slope is an estimate of β and the intercept is an estimate of $log_{10}(\alpha)$. The fit of Equation (7.2) may be used to predict $log_{10}(W)$ given a $log_{10}(L)$, which will then be *back-transformed* (i.e., anti-logged) to the more useful W.

7.3   Fitting Linear Regressions

Simple linear regression is used to model the linear relationship between a single quantitative response (or dependent) variable and a single quantitative explanatory (or independent) variable. This is the case with Equation (7.2) where $log_{10}(W)$ is the quantitative response variable and $log_{10}(L)$ is the quantitative explanatory variable.

Simple linear regressions are fit with lm(). The first argument to lm() is a formula of the form y~x, where y generically represents the response variable and x generically represents the explanatory variable. The data.frame that contains y and x is given to data=. The result of lm() is saved to an object from which specific information will be extracted.

```
> fit1 <- lm(logW~logL,data=ruf90)
```

Estimates of model parameters, sometimes called *coefficients*, are extracted from the lm() object with coef(). Corresponding 95% confidence intervals (CI) are extracted with confint().[5]

```
> coef(fit1)
(Intercept)          logL
  -4.914468    3.025164
> confint(fit1)
                  2.5 %      97.5 %
(Intercept) -5.045255 -4.783680
logL         2.961690   3.088638
```

The intercept parameter is labeled with (Intercept), whereas the slope parameter is labeled with the explanatory variable from the original call to lm(). Thus, for these data, the estimated slope is 3.025 (95% CI: 2.962, 3.089). The estimated intercept is –4.914 (95% CI: –5.045, –4.784).

A significant relationship exists between the response and the explanatory variable if a model with an intercept term and the explanatory variable explains significantly more of the variability in the response variable than a model with just the intercept term. A hypothesis test for comparing these two models is constructed by including the lm() object in Anova() from **car**.[6]

```
> Anova(fit1)
Anova Table (Type II tests)

Response: logW
           Sum Sq  Df  F value    Pr(>F)
logL       16.3104   1  8874.3 < 2.2e-16
Residuals   0.2647 144
```

The large F test-statistic (under "F value") and correspondingly very small p-value (under "Pr(>F)") indicate that the model with both an intercept and $log_{10}(L)$ explains more of the variability in $log_{10}(W)$ than a model with just an intercept. Thus, $log_{10}(L)$ is a significant predictor of $log_{10}(W)$.

A significant relationship between the response and the explanatory variable is also evident if the slope of the linear regression model is significantly different than zero. A hypothesis test for whether the slope is zero or not is obtained, along with other summary information, by including the lm() object in summary().

```
> summary(fit1)

Call:
lm(formula = logW ~ logL, data = ruf90)

Residuals:
     Min        1Q   Median       3Q      Max
-0.11445  -0.02780  0.00245  0.02819  0.13892

Coefficients:
             Estimate Std. Error t value Pr(>|t|)
(Intercept) -4.91447    0.06617  -74.27   <2e-16
logL         3.02516    0.03211   94.20   <2e-16

Residual standard error: 0.04287 on 144 degrees of freedom
Multiple R-squared:  0.984,Adjusted R-squared:  0.9839
F-statistic:  8874 on 1 and 144 DF,  p-value: < 2.2e-16
```

The results under "Coefficients:" contain the parameter estimates (same as from coef()), corresponding standard errors, and the t test statistics and p-values (under Pr(>|t|)) for the two-tailed test that the parameter equals zero. The p-value in the logL row is very small and indicates that the slope is significantly different than zero, which indicates that there is a significant relationship between $log_{10}(W)$ and $log_{10}(L)$.

The top of the output from summary() also includes a repeat of the call to lm() followed by a brief summary of the residuals, which are examined more fully in Section 7.3.3. The bottom of the output contains information about the model fit including an estimate of the variability around the line (i.e., "residual standard error" $=0.0429$) and coefficient of determination (R^2 $=0.984$). The "F-statistic", degrees-of-freedom ("Df"), and "p-value" in the last row are repeated from the `Anova()` results.

7.3.1 Making Predictions

A fisheries scientist may use the fitted model to predict the weight of a fish given its length. For example, the weight-length model derived from a sub-sample of fish for which both weight and lengths were recorded may be used to predict the weight for those fish in the sample for which only length was recorded. Alternatively, the observed weight of a fish may be compared to a predicted weight to determine if the fish has an above or below average weight (see Chapter 8).

Values of the response variable (i.e., $log_{10}(W)$) are predicted from the fitted model using `predict()` with the `lm()` object as the first argument. The second argument is a data.frame that contains values of the explanatory variable (i.e., $log_{10}(L)$) to be used to predict values of the response variable. This data.frame is constructed with `data.frame()`, where the first argument is the **exact same** name as the explanatory variable in the `lm()` object set equal to a vector of values at which to make predictions. Some efficiencies are gained by first putting the untransformed lengths to be used into a vector and then log-transforming that vector and setting it equal to the explanatory variable in a new data.frame. This new data.frame is then the second argument to `predict()`. For example, the code below is used to predict the log weight for 100 and 160 mm Ruffe.

```
> lens <- c(100,160)              # vector of lengths
> nd <- data.frame(logL=log10(lens))  # df of log(lengths)
> ( plogW <- predict(fit1,nd) )   # predicted log(weights)
       1        2
1.135860 1.753356
```

The predicted log weights are back-transformed to the original scale by using the vector of values as the exponent to the base of the logarithm. However, values that are back-transformed from the log scale are commonly biased. A common correction for allometric equations (Sprugel 1983) is to multiply the back-transformed value by

$$e^{\frac{(log_e(base)s_{Y|X})^2}{2}}$$

where $s_{Y|X}$ is the residual standard error ($=0.0429$ from `summary()`) and $log_e(base)$ is used to adjust for the base of the logarithm used.[7] This correction factor is most efficiently computed by supplying the `lm()` object and the base of the logarithm used as the first two arguments to `logbtcf()` from **FSA**.

```
> ( cf <- logbtcf(fit1,10) )  # correction factor
[1] 1.004884
```

The corrected back-transformed predicted value of the response variable is then calculated by multiplying the back-transformed predicted value by this correction factor.

```
> cf*(10^plogW)          # back-transforming with bias correction
        1         2
13.73965 56.94713
```

Thus, this model predicts that the weight of a 100 mm Ruffe is 13.74 g.

Confidence intervals for the **mean** or prediction intervals for an **individual value** of the response variable predicted from the fitted model are obtained by including `interval="confidence"` and `interval="prediction"` in `predict()`, respectively.

```
> mlogW <- predict(fit1,nd,interval="confidence")
> cf*10^mlogW
       fit      lwr      upr
1 13.73965 13.49177 13.99208
2 56.94713 55.43960 58.49566
> plogW <- predict(fit1,nd,interval="prediction")
> cf*10^plogW
       fit      lwr      upr
1 13.73965 11.29455 16.71406
2 56.94713 46.76664 69.34378
```

Thus, one is 95% confident that the **mean** weight of 100 mm Ruffe is between 13.49 and 13.99 g, whereas one is 95% confident that the weight of an **individual** 100 mm Ruffe is between 11.29 and 16.71 g.

7.3.2 Visualizing the Fit

Constructing a plot that shows the fitted line relative to the observed data was demonstrated in Section 3.5.1 for log-transformed and back-transformed weight-length relationships. Similar code is used here to show the model fits for Ruffe captured in 1990 (Figure 7.2).

```
> plot(logW~logL,data=ruf90,pch=19,col=rgb(0,0,0,1/4),
        ylab="log Weight (g)",xlab="log Total Length (mm)")
> tmp <- range(ruf90$logL)
> xs <- seq(tmp[1],tmp[2],length.out=99)
> ys <- predict(fit1,data.frame(logL=xs))
> lines(ys~xs,lwd=2)
```

```
> plot(wt~tl,data=ruf90,pch=19,col=rgb(0,0,0,1/4),
        ylab="Weight (g)",xlab="Total Length (mm)")
> btxs <- 10^xs
> btys <- cf*10^ys
> lines(btys~btxs,lwd=2)
```

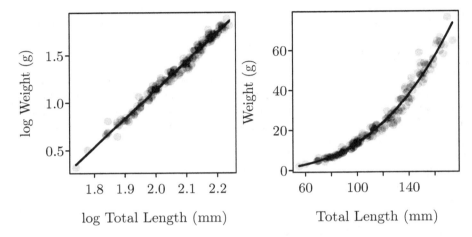

FIGURE 7.2. The log_{10}-log_{10} transformed Ruffe weight-length data with the best-fit line superimposed (Left) and the weight-length data with the back-transformed best-fit line superimposed (Right).

Other methods for adding the best-fit line were mentioned in Section 3.5.1. However, the method using `lines()` is preferred here because it generalizes such that other lines (e.g., confidence or prediction bands) can be added to the plot. For example, `predict()` from above is modified below to include the prediction intervals. In this case, `btys` is a matrix with the corrected back-transformed predicted values in `"fit"` and the corresponding lower and upper prediction bounds in `"lwr"` and `"upr"`, respectively.

```
> btys <- cf*10^predict(fit1,data.frame(logL=xs),
                         interval="prediction")
> head(btys,n=3)
       fit      lwr      upr
1 2.251802 1.841414 2.753652
2 2.333299 1.908384 2.852825
3 2.417746 1.977784 2.955579
```

Each column of `btys` is plotted against `btxs`, as shown below, to construct a plot that includes the best-fit line and lines that represent a 95% prediction band (Figure 7.3).

```
> plot(wt~tl,data=ruf90,pch=19,col=rgb(0,0,0,1/4),
       ylab="Weight (g)",xlab="Total Length (mm)")
> lines(btys[,"fit"]~btxs,col="gray20",lwd=2,lty="solid")
> lines(btys[,"lwr"]~btxs,col="gray20",lwd=2,lty="dashed")
> lines(btys[,"upr"]~btxs,col="gray20",lwd=2,lty="dashed")
```

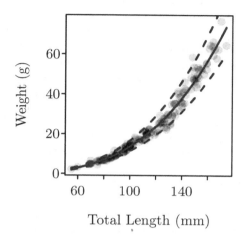

FIGURE 7.3. Ruffe weight-length data with the best-fit model (solid line) and 95% prediction bands (dashed lines) superimposed.

7.3.3 Assumption Checking

The simple linear regression method assumes that a line adequately represents the data and that the measurement errors (i.e., ϵ_i) are independent, are normally distributed, and have a constant variance regardless of the value of the explanatory variable (i.e., homoscedastic).

The independence assumption generally depends on how the data were collected. For example, errors are likely not independent if multiple measurements were made on the same fish over time. Dependent errors are usually not a problem with weight-length data; thus, this assumption will not be addressed further here.[8]

Assessments of the linearity, normality, and constant variance assumptions requires calculation of two related quantities — *fitted values* and *residuals*. Fitted values are values of the response variable predicted from the fitted linear regression model for each individual. Residuals are the difference between the observed value of the response variable and the fitted value for each individual. Individuals with residuals far from zero are not well-represented by the fitted model. Residuals and fitted values are extracted from the lm() object with residuals() and fitted(), respectively.

```
> r <- residuals(fit1)
> fv <- fitted(fit1)
```

Statistical tests exist to determine if the linearity, normality, and homoscedasticity assumptions have been violated. However, the adequacy of these assumptions can often be better assessed by interpreting two plots.[9] A *residual plot* plots residuals against fitted values. The model fit and homoscedasticity assumptions are met if the residual plot exhibits no distinct pattern (Figure 7.4). Curvature in the residual plot indicates that the relationship between the response and explanatory variables is not linear, whereas "funneling" from left to right indicates a nonconstant variance (i.e., heteroscedasticity, Figure 7.4). The normality of measurements assumption is adequately met if a histogram of model residuals is approximately symmetric without overly long "tails."

A residual plot and histogram of residuals (Figure 7.5) are simultaneously constructed by submitting the `lm()` object to `residPlot()` from **FSA**.

```
> residPlot(fit1)
```

In this version, the residual plot is augmented with a loess smoother (with approximate confidence bands) to highlight any curvature in the plot. In addition, any significant outliers[10] are highlighted with the row number of the outlier individual (see Figure 7.4 for examples).

The residual plot shown in Figure 7.5 exhibits linearity because the loess smoother does not vary substantially from a horizontal line (centered on zero). Additionally, the error variance appears constant as there is no obvious funneling of residuals from left to right. Finally, the histogram of residuals in Figure 7.5 is approximately symmetric without extended tails, which indicates approximate normality for the residuals. Thus, the assumptions of simple linear regression are adequately met for these data.

7.4 Among Group Statistical Comparisons

Fisheries scientists often need to determine if the parameters from a simple linear regression model are statistically different between two or more populations. For example, one may want to determine if $log_{10}(\alpha)$ or β in Equation (7.2) differs between sexes, species, lakes, or treatments. These types of questions require fitting a more complicated regression model called a *dummy variable regression* (DVR; Fox 1997) or analysis of covariance (ANCOVA; Pope and Kruse 2007).

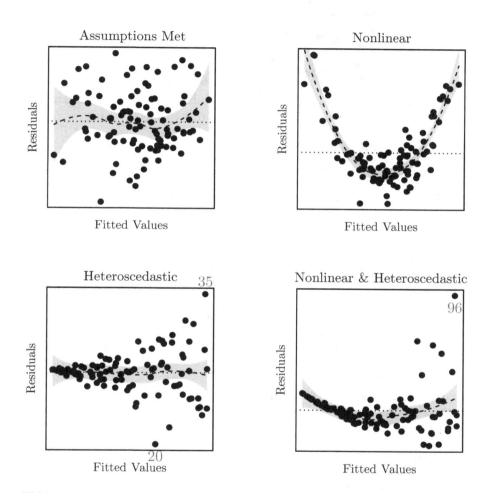

FIGURE 7.4. Modified residual plots that illustrate when the regression assumptions are met (Upper Left) and three common assumption violations. Significant outliers are labeled with the individual number in the bottom two plots.

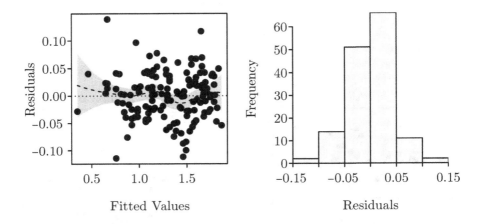

FIGURE 7.5. Modified residual plot (Left) and histogram of residuals (Right) from fitting a linear regression to the log-transformed weights and lengths of Ruffe.

7.4.1 Factor and Dummy Variables

A DVR requires a categorical variable coded as a *factor* (Section 1.7.2) that indicates to which group an individual belongs. In the ruf9000 data.frame, year could be considered as a categorical variable that identifies which of the two years the fish was captured. However, year, recorded as 1990 and 2000, appears numeric to R. Thus, a new variable, called fYear, that forces R to consider 1990 and 2000 as a factor (i.e., group labels) was created with factor() and appended to the data.frame .

```
> ruf9000 %<>% mutate(fYear=factor(year))
> str(ruf9000)
'data.frame': 360 obs. of  7 variables:
 $ year : int   1990 1990 1990 1990 1990 1990 1990 1990 1990 19..
 $ month: int   7 7 7 7 7 7 7 7 7 7 ...
 $ tl   : int   55 60 69 69 70 70 70 75 75 75 ...
 $ wt   : num   2.1 3.2 4.5 4.7 4.8 4.9 6.4 4.4 5.9 6.3 ...
 $ logW : num   0.322 0.505 0.653 0.672 0.681 ...
 $ logL : num   1.74 1.78 1.84 1.84 1.85 ...
 $ fYear: Factor w/ 2 levels "1990","2000": 1 1 1 1 1 1 1 1 1 1 ..
```

A *dummy variable* is a variable that contains the numeric **codes** (not values) created from a factor variable. In the most common coding scheme, a dummy variable consists of zeroes and ones, where a one indicates that the individual has a certain characteristic and a zero indicates that the individual does not have that characteristic. Multiple dummy variables, one fewer than the number of levels of the factor variable, are used when more than two groups are present.

When used in a linear model (see next section), dummy variables are automatically created[11] and are named with the "one" level name appended to the original factor variable name. For example, the dummy variable that corresponds to the `fYear` factor variable in the `ruf9000` data.frame is

$$fYear2000 = \begin{cases} 1, \text{ if captured in } 2000 \\ 0, \text{ if NOT captured in } 2000 \end{cases}$$

7.4.2 Models and Submodels

The DVR model is an extension of Equation (7.2) that includes the quantitative explanatory variable (e.g., $log_{10}(L_i)$), which is often called a *covariate* in this context, as well as the dummy variable(s) and all interactions between the dummy variable(s) and the covariate. Thus, for example, the DVR for when two groups are present will contain three variables — a covariate, one dummy variable, and one interaction between the covariate and the dummy variable.

With these definitions, Equation (7.2) is modified to be

$$\begin{aligned} log_{10}(W_i) = {}& log_{10}(\alpha) + \beta log_{10}(L_i) \\ & + \delta fYear2000 + \gamma fYear2000 * log_{10}(L_i) + \epsilon_i \end{aligned} \tag{7.3}$$

Equation (7.3) simultaneously represents the transformed weight-length relationship for **both** capture years, which is seen by alternately substituting a 0 and a 1 for $fYear2000$ in Equation (7.3) and simplifying to solve for two separate *submodels* as shown in Table 7.1.

An examination of the final submodels (Table 7.1) reveals that α and β are the intercept and slope, respectively, of the transformed weight-length relationship for the fish captured in 1990, and $\alpha+\delta$ and $\beta+\gamma$ are the intercept and slope, respectively, of the transformed weight-length relationship for the fish captured in 2000. Thus, by subtraction, δ is the *difference* in intercepts and

TABLE 7.1. The submodels by capture year represented in Equation (7.3).

Year	$fYear2000$	submodel ($log_{10}(W_i) =$)
1990	0	$= \alpha + \beta log_{10}(L_i) + \delta * 0 + \gamma * 0 * log_{10}(L_i)$ $= \alpha + \beta log_{10}(L_i)$
2000	1	$= \alpha + \beta log_{10}(L_i) + \delta * 1 + \gamma * 1 * log_{10}(L_i)$ $= (\alpha + \delta) + (\beta + \gamma)log_{10}(L_i)$

γ is the *difference* in slopes between fish captured in the two years (Figure 7.6). Thus, as shown in Section 7.4.4, one can test if $\gamma = 0$, or equivalently, whether the interaction term is a significant predictor in the model, to determine if the groups have statistically similar slopes or not.

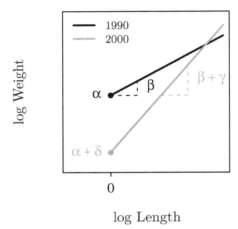

FIGURE 7.6. Representation of the two submodels fit with Equation (7.3). Relative positions of the parameter estimates for each model are shown (note that δ is negative and γ is positive in this depiction).

7.4.3   Model Fitting

Fitting DVR models is easy because R automatically creates the required dummy and interaction variables. The model is fit with `lm()` using `y~x*factor` as the formula, where y generically represents the quantitative response variable, x generically represents the quantitative explanatory variable (i.e., the covariate), and `factor` generically represents the factor explanatory variable. This formula is actually shorthand for the longer but more explicit `y~x+factor+factor:x` formula. Thus, this formula tells R to fit a model with three "effects" or possible explanatory variables — the two main effects of the covariate (x) and the factor (`factor`), and the interaction effect between the covariate and the factor (`factor:x`).

Equation (7.3) is fit to the Ruffe data below, making sure to use the factored version of the year variable (i.e., `fYear`).

```
> fit2 <- lm(logW~logL*fYear,data=ruf9000)
```

The assumptions for a DVR are the same as for a simple linear regression and, thus, assumption checking is conducted as shown previously. The residual plot and histogram of residuals are not shown here; however, the regression assumptions appear to have been adequately met for these data.

7.4.4 Testing Predictors

An understanding of the parameters in Equation (7.3) leads to interesting and important statistical tests. For example, if the interaction term (i.e., $fYear2000*log_{10}(L)$) is not a significant predictor in the model, which implies that $\gamma = 0$, then there is no significant difference in slopes between the two capture years. Additionally, if both the $fYear2000$ and $fYear2000*log_{10}(L)$ variables are not significant predictors in the model, which implies that both $\delta = 0$ and $\gamma = 0$, then neither the slopes nor the intercepts differ significantly between the two years, and, thus, the same line may be used for both years.

In statistical hypothesis testing, it generally does not make sense to test the significance of main effect variables (e.g., $log_{10}(L)$ and $fYear2000$) if the model contains a significant higher-order interaction (Fox and Weisberg 2011). Thus, in the DVR models, the significance of the interaction variable must be determined first. If the interaction is a significant predictor in the model, then no further testing is conducted. However, if the interaction is not a significant predictor in the model, then tests of the main effect variables *ignoring the interaction* may be conducted. Most importantly, hypothesis tests with the main effect variables should occur only if the interaction variable is not significant.

In simple DVR models with one covariate and dummy variables derived from one factor variable, the appropriate tests are conducted and summarized in an *ANOVA table* (Fox and Weisberg 2011) by submitting the `lm()` object to `Anova()` from **car**.[12]

```
> Anova(fit2)
Anova Table (Type II tests)

Response: logW
            Sum Sq  Df   F value     Pr(>F)
logL        32.356   1 15949.981 < 2.2e-16
fYear        0.345   1   170.140 < 2.2e-16
logL:fYear   0.050   1    24.586 1.102e-06
Residuals    0.722 356
```

Following the previous discussion, the results in the ANOVA table should be interpreted sequentially starting from the bottom of the table. The last test indicates whether the interaction variable (`logL:fYear`) is a significant predictor and, thus, indicates whether the groups have the same slope or not. If the slopes are the same, then the middle test indicates whether the dummy variable (`fYear`) is a significant predictor and, thus, whether the groups have the same intercept or not. If the slopes are the same, then the top test indicates whether the covariate (`logL`) is a significant predictor and, thus, whether the groups have a significant (shared) slope or not. If the bottom test suggests that the slopes are significantly different, then the top two tests are generally meaningless and should not be interpreted.[13]

These results suggest that Ruffe captured in 1990 and those captured in 2000 have significantly different slopes (F=24.6, $p < 0.00005$). Thus, the weight-length relationship for Ruffe should be modeled with different equations for 1990 and 2000.

Parameter estimates and confidence intervals are extracted with `coef()` and `confint()`, respectively. Below, the results are column-bound together with `cbind()` for a more succinct presentation.

```
> cbind(coef=coef(fit2),confint(fit2))
                      coef      2.5 %        97.5 %
(Intercept)    -4.9144676 -5.051181 -4.7777541
logL            3.0251636  2.958814  3.0915136
fYear2000       0.3942964  0.210879  0.5777138
logL:fYear2000 -0.2285159 -0.319152 -0.1378797
```

From this, it is seen that the slope for the 2000 capture year (γ) is 0.229 (95% CI: 0.319, 0.138) **lower** than the slope for the 1990 capture year. Additionally, the slope for the 1990 capture year (β) is 3.025 (95% CI: 2.959, 3.092).[14]

7.4.5 Visualizing the Fit

A visual of the two weight-length relationships (Figure 7.7) may be constructed using the methods described in previous sections and in Section 3.1.2.

```
> # plot points with different symbols for each year
> symbs <- c(1,16)
> plot(wt~tl,data=ruf9000,pch=symbs[ruf9000$fYear],
        ylab="Weight (g)",xlab="Total Length (mm)",cex=0.6)
> # find range of tl values in each year
> tmp <- ruf9000 %>% group_by(fYear) %>%
      summarize(min=min(tl,na.rm=TRUE),
                max=max(tl,na.rm=TRUE))
> # plot line for 1990
> tmpx <- seq(tmp$min[1],tmp$max[1],length.out=99)
> tmpy <- 10^(predict(fit2,
               data.frame(logL=log10(tmpx),fYear=factor(1990))))
> lines(tmpy~tmpx,col="gray60",lwd=2)
> # plot line for 2000
> tmpx <- seq(tmp$min[2],tmp$max[2],length.out=99)
> tmpy <- 10^(predict(fit2,
               data.frame(logL=log10(tmpx),fYear=factor(2000))))
> lines(tmpy~tmpx,col="gray90",lwd=2)
> # add a legend
> legend("topleft",c("1990","2000"),pch=symbs,cex=0.9,bty="n")
```

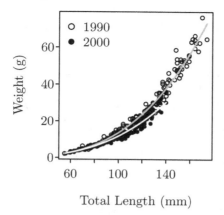

Total Length (mm)

FIGURE 7.7. Comparison of weight-length model fits for Ruffe captured in 1990 and 2000.

In the case where the slopes of the $log_{10}(W)$-$log_{10}(L)$ relationship are the same between groups, then the difference in intercepts is a measure of the difference in mean $log_{10}(W)$ between groups at all values of $log_{10}(L)$. In other words, the intercept is the constant difference in mean $log_{10}(W)$ between groups no matter the value of $log_{10}(L)$. In this case, mean W for one group is a constant *multiple* of the mean W for the other group at the same L.

However, when the slopes differ between groups, the difference in intercepts is only a measure of how the mean $log_{10}(W)$ differs between groups when $log_{10}(L) = 0$, which is, at best, of limited interest. A simple interpretation regarding differences in mean $log_{10}(W)$ between groups is impossible when the slopes differ because the mean $log_{10}(W)$ might differ for some values of $log_{10}(L)$ but not for others.

Predicting the mean W, along with appropriate confidence intervals, for both groups for a variety of values of L can aid interpretation when the slope of the weight-length relationship differs between groups. A visual of predicted weights is constructed with lwCompPreds() from **FSA**. The first argument to lwCompPreds() is the lm() object that was constructed to compare the groups. By default, lwCompPreds() will predict W at the 5th, 25th, 50th, 75th, and 95th quantiles (also called percentiles) of observed lengths in the sample. Different quantiles are chosen by providing a vector of quantiles to qlens=. Alternatively, specific lengths may be used by providing a vector of lengths to lens=. Finally, the base of the logarithm used must be included in base=, unless natural logarithms were used (i.e., the default).

An example of predicted weights at the five default quantile lengths of Ruffe captured in 1990 and 2000 is shown in Figure 7.8. These results suggest that difference in mean weight between years increases considerably as the length of Ruffe increases.

```
> lwCompPreds(fit2,base=10,xlab="Year",main.pre="TL=")
```

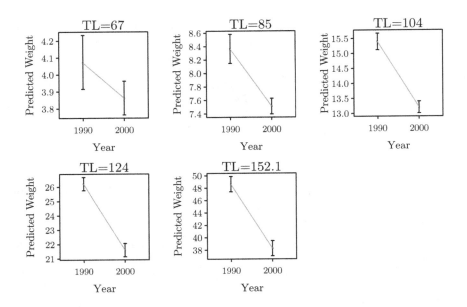

FIGURE 7.8. Predicted weights, with 95% confidence intervals, at the 5th, 25th, 50th, 75th, and 95th percentiles of total lengths (TL) for Ruffe captured in July in years from 1990 and 2000.

This method is more useful when many groups (e.g., capture years) are considered (Figure 7.9).

7.5   Further Considerations

Other types of regression models have been fit to weight-length data. Nonlinear methods (see Section 12.3) have been used in situations where the errors are additive rather than multiplicative around the power function relationship. Nonlinear mixed effects models (e.g., Zuur et al. 2009) may also be useful for fitting the power function and modeling the error structure without transformation. Several authors, largely following the related work of Gerow et al. (2005), fit a quadratic model to the $log_{10}(W)$-$log_{10}(L)$ relationship to account for a nonlinear form still inherent after transformation. Jellyman et al. (2013) suggested that the quadratic model may be appropriate for species that grow to very large sizes. Quantile regression (Cade and Noon 2003) has been used

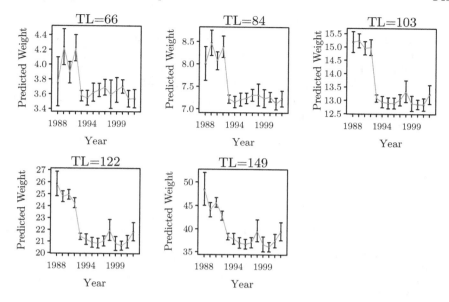

FIGURE 7.9. Predicted weights, with 95% confidence intervals, at the 5th, 25th, 50th, 75th, and 95th percentiles of total lengths (TL) for Ruffe captured in July in years from 1988 to 2002.

to model aspects of the weight-length relationship other than the central tendency (i.e., mean). Cade et al. (2011) provided an interesting case study for the use of quantile regression with weight-length data.

The most common "other" method for analyzing weight-length data is the *geometric mean regression*[15] method argued for by Ricker (1973). The basis of the argument for this method is that the explanatory variable in the ordinary least-squares methods used in this chapter is assumed to be measured without error. The geometric mean regression method accounts for errors in the explanatory variable. These methods are implemented in a number of packages in R, including `Deming()` from **MethComp**, `gx.rma()` from **rgr** (Garrett 2015), and `odregress()` from **pracma** (Borchers 2015). Use of the geometric mean regression has not gained favor in the analysis of weight-length relationships, likely because of historical precedence and inertia, there is little measurement error in fish lengths, or there is little impact on predicted values for many species.

The methods for comparing two relationships described in Section 7.4 are easily extended to more than two groups (see the online supplement). However, a test of the interaction variable only suggests that at least one group has a different slope from the other groups. A multiple comparison procedure to determine which groups have different slopes is not readily available. All pairwise comparisons can be made, however, and the vector of p-values submitted to `p.adjust()` to control the experimentwise error rate. More complex statistical models, including mixed models (Lai and Helser 2004) and hierarchical Bayesian models (He et al. 2008) also provide methods for comparing the weight-length relationships among multiple groups.

Finally, `lmList()` and related functions from **lme4** (Bates et al. 2014) provide a quick way to compute (without comparing) weight-length relationship parameters for a large number of groups.

Notes

[1] More thorough introductions to simple linear and dummy variable regressions are found in any number of general statistics, method-specific statistics (e.g., Fox and Weisberg 2011; Weisberg 2014), or fisheries-specific (e.g., Guy and Brown 2007) books.

[2] Data manipulations in this chapter require functions from **magrittr** and **dplyr**, which are fully described in Chapter 2.

[3] If the variability around the line does not increase with increasing length of the fish, then the error term is likely not multiplicative and is, rather, additive. In this case, the weight-length relationship should be examined with the nonlinear model described in Section 12.3.

[4] Multiplicative errors may be modeled as any constant raised to the power of the ϵ_i. It is fairly typical with fisheries data to use common logarithms to linearize the power function and, thus, the constant of 10 is used here. If e had been used as a constant, then the natural logarithms would have been used and if 2 had been the constant then the logarithm to base 2 would have been used. The use of different bases will result in different estimates of the y-intercept, but the slope, all predictions, and all hypothesis tests will be the same. Thus, the base used in analyzing weight-length data is irrelevant, except that the analyst must be aware of which base is being used so that the data are properly transformed and back-transformed.

[5] The level of confidence for the confidence interval can be changed with `conf.level=`.

[6] `Anova()` from **car** computes Type II sums-of-squares (SS) and tests. In simple linear regression, Type II and Type I SS and tests are equivalent. The `anova()` function from base R returns Type I SS and tests and, thus, can also be used here. However, `Anova()` is used throughout this book. Type I and Type II SS are discussed in more detail in Section 7.4.

[7] The method developed by Sprugel (1983) used natural logarithms. However, he also provided this adjustment for when a different base was used.

[8] Analysis of data with dependent errors requires use of other statistical models that account for the structure of the dependency (i.e., repeated measures and mixed models; e.g., Faraway 2005; Zuur et al. 2009).

[9] Fox and Weisberg (2011) provided a thorough treatment of assumption checking for linear regression models.

[10] Significant outliers are identified with a Bonferroni-adjusted t-test of the externally Studentized residuals as implemented in `outlierTest()` from **car**.

[11] When dummy variables are created, the first level (or group) in the factor variable is coded with a 0.

[12] Type I SS are sometimes called *sequential SS* because the SS (and subsequent tests) are computed for each term in the model after considering all previous terms in the model. Thus, the SS for the interaction term (last term in the model) are computed after considering both main effects. However, the SS for the second main effect are computed without considering the interaction term and the SS for the first main effect are computed without considering either the interaction term or the other main effect. The order of the

two main effects is irrelevant because they are uncorrelated in this case, where one is a covariate and the other is a factor.

Type II SS follow the *marginality principle* (Fox and Weisberg 2011) where the SS for one term are adjusted for all other terms in the model except for related higher-order interactions. In the simple DVR, the interaction term is adjusted for the two main effects, but each main effect is adjusted for the other main effect but not for the interaction term. Because the main effects are uncorrelated, adjusting the SS of one main effect for the other main effect is inconsequential. Thus, for the simple DVR model presented here, the Type I and Type II SS are equivalent.

Type II SS are used throughout the book. See Fox and Weisberg (2011) and Aho (2014) for further discussion of the types of SS.

[13]For example, if the interaction term is not significant, then there is no difference in slopes between the groups (i.e., lines are parallel). In this case, further testing of the covariate main effect indicates whether or not there is a significant relationship (i.e., slope) between the response and explanatory variable that is shared by both groups. Additionally, further testing of the factor main effect determines if the two groups have the same intercept or not. This latter test is important because if the intercepts differ between groups that have the same slope, then that difference is maintained at all values of the covariate, not just where the covariate is equal to zero. Alternatively, if the interaction term is significant, then the groups have significantly different slopes. Testing the covariate effect does not make sense because the relationship between the response and explanatory variable differs between the groups. Similarly, testing the factor main effect does not make sense because, with different slopes between groups, a difference in intercepts does not generalize to any other values of the covariate.

[14]There is no simple way, from the results where models are fit to both years simultaneously, to construct a confidence interval for either the slope or intercept of Ruffe captured in 2000. Thus, confidence intervals for the parameters for each group are usually computed by creating data.frames for each group, fitting the model to each group separately, and then extracting the desired confidence intervals.

[15]Synonymous or slightly variant methods to geometric mean regression include error-in-variables, Deming, model II, orthogonal, reduced major axis, or measurement error regression.

8

Condition

Condition is a measure of the physical health of a population of fish based on the relative heaviness or plumpness of fish in that population. The condition of an individual fish is most often computed by comparing the observed weight of the fish to an expected weight based on the fish's observed length. In other words, the condition of a particular fish is a matter of determining if it weighs more or less than would be expected based on its length. An overall measure of condition for an entire population is the average or median condition of all fish in a sample.

The utility of measuring fish condition was summarized by Blackwell et al. (2000, p. 2):

> Fish condition can be extremely important to fisheries managers. Plump fish may be indicators of favorable environmental conditions (e.g., habitat conditions, ample prey availability), whereas thin fish may indicate less favorable environmental conditions. Thus, being able to monitor fish well-being can be extremely useful for fisheries biologists who must make management recommendations concerning fish populations.

There are at least eight metrics of condition (Bolger and Connolly 1989). Four of these measures are commonly used by fisheries scientists (Blackwell et al. 2000) and are described in Section 8.2. The computation of the condition metrics for each fish in a sample is simply a matter of adding a variable to a data.frame (see Section 2.2.5). Methods to summarize and compare overall condition among groups of fish are more involved and are shown in Section 8.3.

Required Packages for This Chapter

Functions used in this chapter require loading the packages shown below.[1]

```
> library(FSA)
> library(car)          # Before dplyr to reduce conflicts with MASS
> library(magrittr)
> library(dplyr)
> library(plotrix)
> library(multcomp)
```

8.1   Data Requirements

The same data used to describe a weight-length relationship (Section 7.1) are used to assess fish condition. Thus, the same data used in Section 7.1 — weights (g) and total lengths (mm) of Ruffe captured in the St. Louis River Harbor (Minnesota and Wisconsin) in July — are used here. Here, however, Ruffe captured in 1990, 1995, and 2000 are examined, and several variables are removed to save space.

```
> ruf950 <- read.csv("RuffeSLRH.csv") %>%
    filter(month==7,year %in% c(1990,1995,2000)) %>%
    mutate(logW=log10(wt),logL=log10(tl)) %>%
    select(-c(fishID,month,day))
> headtail(ruf950,n=2)
    year  tl   wt       logW      logL
1   1990  55   2.1  0.3222193  1.740363
2   1990  60   3.2  0.5051500  1.778151
502 2000 140  30.2  1.4800069  2.146128
503 2000 140  36.2  1.5587086  2.146128
```

8.2   Condition Metrics

The four most common measures of fish condition are *Fulton's condition factor*, *weight-length relationship residuals*, *relative condition*, and *relative weight*. These metrics are introduced below, but are described in more detail in Bolger and Connolly (1989), Pope and Kruse (2007), Neumann et al. (2012), and Lloret et al. (2014).

8.2.1   Fulton's Condition Factor

Fulton's condition factor (K_i) for an individual fish is

$$K_i = \frac{W_i}{L_i^3} * constant$$

where L_i and W_i are the observed length and weight for the ith fish and the constant is a scaling factor equal to $100,000$ if metric units (i.e., grams and millimeters) were recorded or $10,000$ if English units (i.e., pounds and inches) were recorded (Neumann et al. 2012).[2] Fulton's condition factor *assumes isometric growth* where the shape of the fish does not change with increasing length (Lloret et al. 2014). If the population of fish being examined does not exhibit isometric growth, which is often the case (Froese 2006), then K_i depends on the length of the fish (Lloret et al. 2014). Thus, comparisons of Fulton's condition factor are restricted to fish of similar lengths within the population (Pope and Kruse 2007; Lloret et al. 2014). Fulton's condition factor is seldom used in modern studies because of this limitation.

The K_i are added to the data.frame below.

```
> ruf950 %<>% mutate(K=wt/(tl^3)*100000)
> headtail(ruf950,n=2)
    year  tl  wt       logW      logL        K
1   1990  55  2.1 0.3222193 1.740363 1.262209
2   1990  60  3.2 0.5051500 1.778151 1.481481
502 2000 140 30.2 1.4800069 2.146128 1.100583
503 2000 140 36.2 1.5587086 2.146128 1.319242
```

8.2.2 Weight-Length Residuals

Residuals from the weight-length relationship (Chapter 7) have also been used to measure fish condition (Sutton et al. 2000; Pope and Kruse 2007). The residual for the ith fish (e_i) is the difference between the observed log weight and the log weight predicted from substituting the fish's observed log length into the (log) weight-length regression, or

$$e_i = log_{10}(W_i) - (log_{10}(a) + blog_{10}(L_i))$$

For example, a negative residual indicates that the log weight is less than average given the corresponding log length; thus, that fish would be in relatively poorer condition than an average fish of that log length in the sample.

The weight-length relationship for the sample of Ruffe is computed with lm() below (see Section 7.3 for more details).

```
> lm1 <- lm(logW~logL,data=ruf950)
> coef(lm1)
(Intercept)        logL
  -4.857360    2.975671
```

The residuals are extracted from the lm() object with **residuals()** and appended to the data.frame below.

```
> ruf950 %<>% mutate(lwresid=residuals(lm1))
> headtail(ruf950,n=2)
    year  tl  wt       logW      logL        K      lwresid
1   1990  55  2.1 0.3222193 1.740363 1.262209  0.000833274
2   1990  60  3.2 0.5051500 1.778151 1.481481  0.071317647
502 2000 140 30.2 1.4800069 2.146128 1.100583 -0.048803103
503 2000 140 36.2 1.5587086 2.146128 1.319242  0.029898525
```

From this it is seen that fish #1 has a slightly above average and fish #502 has a below average log weight for its log length.

8.2.3 Relative Condition Factor

The relative condition factor (Kn_i; Le Cren 1951) for the ith fish is

$$Kn_i = \frac{W_i}{\widehat{W_i}}$$

where $\widehat{W_i}$ is the predicted mean weight given the fish's observed length for the population of fish under investigation (Blackwell et al. 2000). The value of Kn_i does not depend on length within the population being investigated (Neumann et al. 2012). However, Bolger and Connolly (1989) showed that Kn_i can only be compared among groups of fish that have the same slope for the weight-length relationship (i.e., same b in Equation (7.2)).[3] Thus, $\widehat{W_i}$ is generally predicted from a weight-length equation for all groups of interest that have the same slope for the weight-length relationship (perhaps from several years, locations, or regions; Lloret et al. 2014).

The predicted log weights given the log length of each fish are extracted from the `lm()` object with `fitted()`. Using these values as the exponent of 10 (i.e., anti-logging) returns the $\widehat{W_i}$.[4] The Kn_i are then added to the data.frame by dividing the observed weights by the $\widehat{W_i}$.

```
> ruf950 %<>% mutate(predW=10^fitted(lm1),Kn=wt/predW)
> headtail(ruf950,n=2)
    year  tl   wt      logW     logL        K     lwresid
1   1990  55  2.1 0.3222193 1.740363 1.262209 0.000833274
2   1990  60  3.2 0.5051500 1.778151 1.481481 0.071317647
502 2000 140 30.2 1.4800069 2.146128 1.100583 -0.048803103
503 2000 140 36.2 1.5587086 2.146128 1.319242 0.029898525
       predW        Kn
1   2.095975 1.0019205
2   2.715391 1.1784676
502 33.791700 0.8937106
503 33.791700 1.0712690
```

Values of $Kn_i < 1$ indicate that the fish's weight is below average for fish of the same length in the given population. For example, these results for Ruffe show that fish #1 has a slightly above average and fish #502 has a below average weight for its length.

The Kn_i are equal to the back-transformed residuals from the weight-length relationship (i.e., the correlation (computed with `cor()`) between the Kn_i and back-transformed residuals is 1).

```
> cor(ruf950$Kn,10^ruf950$lwresid)
[1] 1
```

Thus, in a relative sense, the weight-length residuals and the relative condition values relate the same information about the condition of the fish (Peig and Green 2010).[5]

8.2.4 Relative Weight

The relative weight metric has become the most popular measure of condition (Blackwell et al. 2000; Pope and Kruse 2007). The relative weight (Wege and Anderson 1978) for an individual fish is

$$Wr_i = \frac{W_i}{Ws_i} * 100 \qquad (8.1)$$

where Ws_i is a "standard weight" for fish of the same observed length. Depending on the species, the standard weight is either computed through back-transformation of the common log-transformed linear model (Section 7.2)

$$Ws_i = 10^{\alpha + \beta log_{10}(L_i)} \qquad (8.2)$$

or through back-transformation of the log-transformed quadratic model

$$Ws_i = 10^{\alpha + \beta_1 log_{10}(L_i) + \beta_2 (log_{10}(L_i))^2} \qquad (8.3)$$

Standard weight equations (i.e., values for α, β, β_1, and β_2 in Equations (8.2) and (8.3)) have been developed for many common game species in North America and, recently, for some nongame species in North America and other species outside of North America. The equations for computing standard weights are species-specific, are in both metric (mm and grams) and English (inches and pounds) units, use common rather than natural logarithms[6], and were derived from weight-length relationships from throughout the geographical range of a species. Most standard weight equations are linear and were developed such that the standard is the 75th percentile of mean weights at a given length from throughout the geographic range of the species. However, some recent equations have included a quadratic term or were developed for different percentiles (Ogle and Winfield 2009; Cooney and Kwak 2010).

Coefficients for accepted standard weight equations (on the log_{10}-transformed scale) for many species are available in Neumann et al. (2012). In addition, these coefficients are returned by wsVal() from **FSA** with the species name in quotes as the first argument.[7] The default is to return the coefficients for metric units and for the equation derived for the 75th percentile values. Coefficients in English units are returned by including units="English" and other percentiles, *if they exist*, are returned by giving the percentile to ref=.

```
> wsVal("Ruffe")
    species  units       type ref measure method min.TL max.TL
131   Ruffe metric quadratic  75      TL    EmP     55    205
        int slope   quad                     source
131 -2.58 0.621 0.6073 Ogle and Winfield (2009)
                              comment
131 not in Neumann et al. (2012)
```

The first six items returned by wsVal() are reminders of the species; whether metric or English units are reported; whether the equation is linear or quadratic; whether the reference percentile is 25, 50, or 75; the type of length measure reported (TL is total length and FL is fork length); and the type of method used to construct the standard weight equation.[8] The last two returned items show the primary source and any comments for the equation. Standard weight equations are only applicable above some minimum length and, for some species, below a maximum length. Values for the range of acceptable lengths are given under min.XX and max.XX, with the XX replaced by the type of length measurement. Finally, the coefficients for the intercept, slope, and quadratic (if it exists) terms are under int, slope, and quad, respectively.[9] A subset of the most useful of these results is returned if simplify=TRUE is used in wsVal().

```
> ( wsRuf <- wsVal("Ruffe",simplify=TRUE) )
    species min.TL max.TL   int slope   quad
131   Ruffe     55    205 -2.58 0.621 0.6073
```

The coefficients for the standard weight equation are extracted from the wsVal() object below.[10]

```
> # Below for demonstration purposes
> wsRuf[["int"]]
[1] -2.58
> wsRuf[["slope"]]
[1] 0.621
> wsRuf[["quad"]]
[1] 0.6073
```

These values are then used to compute, using Equation (8.2) or Equation (8.3), and append standard and relative weights to the data.frame.[11] In this example, the standard weight equation for Ruffe contains a quadratic term such that Equation (8.3) is used. Most standard weight equations do not have a quadratic term and will use Equation (8.2) (i.e., the quadratic portion of the code below would be omitted for most species).

```
> ruf950 %<>% mutate(Ws=10^(wsRuf[["int"]]+wsRuf[["slope"]]*logL
                          +wsRuf[["quad"]]*(logL^2)),
                  Wr=wt/Ws*100)
> headtail(ruf950,n=2)
    year  tl   wt      logW     logL        K      lwresid
1   1990  55  2.1 0.3222193 1.740363 1.262209  0.000833274
2   1990  60  3.2 0.5051500 1.778151 1.481481  0.071317647
502 2000 140 30.2 1.4800069 2.146128 1.100583 -0.048803103
503 2000 140 36.2 1.5587086 2.146128 1.319242  0.029898525
        predW        Kn       Ws        Wr
1    2.095975 1.0019205 2.188736  95.94581
2    2.715391 1.1784676 2.782315 115.01213
502 33.791700 0.8937106 35.472162 85.13719
503 33.791700 1.0712690 35.472162 102.05186
```

Values of $Wr_i < 100$ indicate that the fish's weight is below the standard weight for fish of the same length. For example, these results indicate that the weights for fish #1 and fish #502 are less than the *75th percentile of mean weights* for Ruffe of the same length. In contrast to the previous two metrics, fish #1 is below the "standard" here, but the standard is much different in the relative weight than in the residuals and relative condition metrics.[12]

8.3 Among Group Statistical Comparisons

The fisheries scientist is usually interested in a summary of condition for a sample or for groups of individuals within a sample rather than for an individual fish. For example, the scientist may be interested in determining if mean condition differed among fish captured in different years, between habitats or waterbodies, or among different length categories. In this section, two methods to compare the mean or median relative weight among different groups of fish are illustrated. The relative weight metric is used in these examples, but any of the condition metrics from Section 8.2 could be used.

For the examples below, all Ruffe shorter than the minimum length and longer than the maximum length for which the standard weight equation should be applied were removed, a variable with the Gabelhouse length categories (see Section 6.3.1) was added, and a new data.frame that contained only Ruffe caught in 1990 was created.

```
> ruf90 <- ruf950 %>%
    filter(tl>=wsRuf[["min.TL"]],tl<=wsRuf[["max.TL"]]) %>%
    mutate(gcat=lencat(tl,breaks=psdVal("Ruffe"),
                      use.names=TRUE)) %>%
    filterD(year==1990)
```

The goal of the ensuing analyses is to determine if the mean or median relative weight of Ruffe differed among the four Gabelhouse length categories.

8.3.1   Summary Statistics

Summary statistics by group can be computed in a number of ways, including the use of group_by() and summarize() from **dplyr** as illustrated in Section 2.5. One other method is to use Summarize() from **FSA**. The first argument to this function is a formula of the form quant~factor1 or quant~factor1*factor2 with quant generically representing a quantitative variable (e.g., relative weight) and factor1 and factor2 generically representing group factor variables (e.g., length category or capture year). The data.frame that contains the variables in the formula is given in data= and the number of decimals to be reported is controlled with digits=.

```
> Summarize(Wr~gcat,data=ruf90,digits=0)
       gcat  n nvalid mean sd min  Q1 median  Q3 max percZero
1     stock 29     29  109 13  84 101    107 114 149        0
2   quality 47     47  112 10  92 105    113 120 132        0
3 preferred 31     31  105 11  84  96    108 112 125        0
4 memorable 39     39  107 10  89 100    106 113 138        0
```

8.3.2   One-Way ANOVA

A *one-way analysis of variance* (ANOVA) is a statistical method to determine if the means from two or more samples were drawn from the same population (i.e., the means are equal). For example, a one-way ANOVA may be used to determine if the mean relative weight of Ruffe differs among the four Gabelhouse length categories. The null hypothesis of this test is that the means from all groups are equal and the alternative hypothesis is that the mean for at least one group differs from the mean of one or more of the other groups. The mathematical theory and statistical assumptions that form the foundation of the one-way ANOVA are described in nearly all introductory statistics books.[13] Methods for assessing the assumptions of a one-way ANOVA are discussed in Section 8.3.2.1.

A one-way ANOVA is conducted using the same lm() function used to fit a linear regression (Section 7.3). However, the right-hand side of the formula must be a factor variable to perform the ANOVA (i.e., the formula must be of the form y~factor).

```
> aov1 <- lm(Wr~gcat,data=ruf90)
```

The results, including the F test-statistic and corresponding p-value, from a one-way ANOVA are usually presented in an *ANOVA table* (Fox and Weisberg 2011). The ANOVA table is extracted from the `lm()` object with `Anova()` from **car**.[14]

```
> Anova(aov1)
Anova Table (Type II tests)

Response: Wr
           Sum Sq  Df F value Pr(>F)
gcat       1131.7   3  3.3019 0.0222
Residuals 16223.2 142
```

The small p-value ($p = 0.022$) in these results indicates that the mean from at least one group is different from the mean of one or more of the other groups. The one-way ANOVA, unfortunately, does not tell which means differ.

A variety of follow-up procedures have been defined in the statistical literature to determine specifically which groups have different means. Most of these methods are designed to control the *experimentwise error rate* at a nominal level chosen by the scientist.[15] *Tukey's Honestly Significant Difference* (HSD) method (Aho 2014) is, perhaps, the most common of these methods.

Tukey's procedure is computed with `glht()` from **multcomp** (Hothorn et al. 2008).[16] This function requires the `lm()` object as the first argument and the `mcp()` function, also from **multcomp**, as the second argument. The argument to `mcp()` is the name of the factor variable from the `lm()` object set equal to `"Tukey"`. The result should be saved to an object that is then submitted to `summary()` to extract the results for all pairwise hypothesis tests.[17]

```
> mc1 <- glht(aov1,mcp(gcat="Tukey"))
> summary(mc1)
                         Estimate Std. Error t value p value
quality - stock == 0        3.303      2.524   1.308   0.558
preferred - stock == 0     -3.900      2.761  -1.412   0.493
memorable - stock == 0     -2.035      2.621  -0.777   0.864
preferred - quality == 0   -7.202      2.473  -2.912   0.021
memorable - quality == 0   -5.338      2.315  -2.306   0.101
memorable - preferred == 0  1.864      2.572   0.725   0.887
```

These results are used to determine whether the difference in means for each pair of groups is significantly different from zero. The order in which the difference in means is computed is shown in the far left column of the resulting table. The p-values for each paired comparison are found in the column labeled with "Pr(>|t|)". The only significant p-value in these results is for the comparison of the preferred- and quality-length groups ($p = 0.0210$). Thus, it appears that the mean relative weight for quality-length Ruffe is significantly greater than that for preferred-length Ruffe (because the mean of the quality-length fish subtracted from the mean of the preferred-length fish is negative).

A two-step process is required to construct a plot of the means for each group with corresponding confidence intervals (Figure 8.1). The first step in this process is to create the results to be plotted. A data.frame of the group names (i.e., length categories) is constructed first with factor() and data.frame(). Note the use of levels= in factor() to force R not to reorder the groups alphabetically. Means (i.e., predicted values) are then computed with predict() using the one-way ANOVA lm() object as the first argument, the data.frame of group names as the second argument, and predict="confidence" to produce the corresponding confidence intervals.

```
> grps <- c("stock","quality","preferred","memorable")
> nd <- data.frame(gcat=factor(grps,levels=grps))
> ( pred <- predict(aov1,nd,interval="confidence") )
       fit      lwr      upr
1 108.9411 105.0174 112.8647
2 112.2436 109.1616 115.3257
3 105.0415 101.2465 108.8365
4 106.9057 103.5223 110.2892
```

The second step in constructing this plot uses plotCI() from **plotrix** to plot the means as a point and confidence intervals as a vertical bar for each group as illustrated in Section 3.5.2. In the example below, the plotting of the x-axis labels was suppressed with xaxt="n" so that labels for only the groups can be added with axis(). A line to connect the means was added with lines(). Use of axis() and lines() was illustrated in Chapter 3.

```
> plotCI(as.numeric(nd$gcat),pred[,"fit"],
         li=pred[,"lwr"],ui=pred[,"upr"],
         pch=19,xaxt="n",xlim=c(0.8,4.2),ylim=c(100,120),
         xlab="Gabelhouse Length Category",ylab="Mean Wr")
> lines(nd$gcat,pred[,"fit"],col="gray50")
> axis(1,at=nd$gcat,labels=nd$gcat)
```

Finally, each point on this plot may be augmented with a *significance letter*, where two points that have the same letter have statistically equal means and those with different letters have statistically different means. The significance letters are extracted from the glht() object with cld() from **multcomp**.[18]

```
> cld(mc1)
    stock   quality preferred memorable
     "ab"       "b"       "a"      "ab"
```

The significance letters are added above the upper-bound of the confidence interval using text() as described in Section 3.6.2. Make sure to enter the letters in the same order that the groups are plotted.

```
> text(x=nd$gcat,y=pred[,"upr"],
        labels=c("ab","b","a","ab"),pos=3)
```

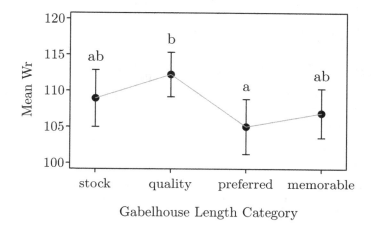

Gabelhouse Length Category

FIGURE 8.1. Mean relative weight by Gabelhouse length category for Ruffe captured in July 1990. Vertical lines indicate 95% confidence intervals for the mean. Means with a common letter are not statistically different.

8.3.2.1 Assumptions

The adequacy of the one-way ANOVA method relies on the meeting of three assumptions (Aho 2014) — the measurement errors (i.e., ϵ_i) are independent, normally distributed in each group, and have a constant variance in each group (i.e., homoscedastic). If these assumptions are not met, then other methods (see Section 8.3.3) should be considered.

As with linear regression, the adequacy of the normality and heteroscedasticity assumptions may be assessed by interpreting the residual plot and the histogram of residuals (Figure 8.2), both of which are produced by `residPlot()` with the `lm()` object.

```
> residPlot(aov1)
```

The residual plot for a one-way ANOVA model is shown as boxplots of the residuals in each group (Figure 8.2-Left). This plot is somewhat difficult to interpret; however, if the height of the boxes are roughly equivalent, then the homoscedasticity assumption is likely met. A simple hypothesis test, however, for testing equal variances among groups is the Levene's test. The null hypothesis for Levene's test is that the variances in all groups are equal, whereas the alternative hypothesis is that variance of at least one group differs from one or

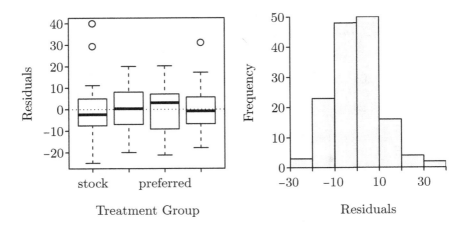

FIGURE 8.2. Residual plot (Left) and histogram of residuals (Right) from fitting a one-way ANOVA to relative weight by Gabelhouse length category for Ruffe captured in 1990.

more of the other groups. Levene's test is conducted by submitting the `lm()` object to `leveneTest()` from **car**.

```
> leveneTest(aov1)
Levene's Test for Homogeneity of Variance (center = median)
       Df F value Pr(>F)
group   3  0.4563 0.7133
       142
```

This large p-value ($p = 0.7133$) indicates that the variances are probably equal and that the homoscedasticity assumption is likely met.

The validity of the one-way ANOVA assumptions when assessing differences in mean relative weights among groups has received considerable attention. Brenden et al. (2003) showed that the Wr_i values are not independent and identically distributed. Relative weight data are often heteroscedastic, which is a violation of the assumption of equal variances among groups. Violations of the homoscedasticity assumption may be rectified by transforming data (Aho 2014), but transformations are often not effective with relative weight data (Murphy et al. 1990; Brenden et al. 2003). Finally, the distribution of Wr_i values may be nonnormal (typically leptokurtic; Pope and Kruse 2007), though they are usually symmetric (Murphy et al. 1990). The one-way ANOVA is robust to slight departures from normality if homoscedasticity is met and the group sample sizes are equal (Quinn and Keough 2002).

8.3.3 Kruskal-Wallis Test

The most common nonparametric equivalent to the one-way ANOVA is the Kruskal-Wallis test. The Kruskal-Wallis test will test, if the distributions are similarly shaped with equal variances among groups, that the medians are equal among all groups (Aho 2014; Hollander et al. 2014). The Kruskal-Wallis test is conducted with `kruskal.test()` using the same arguments as used in `lm()` for the one-way ANOVA.

```
> kruskal.test(Wr~gcat,data=ruf90)

    Kruskal-Wallis rank sum test

data:  Wr by gcat
Kruskal-Wallis chi-squared = 9.9195, df = 3, p-value = 0.01926
```

This small p-value ($p = 0.0193$) indicates that the median for at least one group differs from the medians of one or more of the other groups.

Dunn (1964) described a multiple comparison procedure for the Kruskal-Wallis test, which is implemented in `dunnTest()` from **FSA**.[19] The first two arguments to `dunnTest()` are the same as those to `kruskal.test()`. The p-values produced by the Dunn method must be adjusted to control the experimentwise error rate. The default method used by `dunnTest()` to control the experimentwise error rate is the Bonferroni method, which is highly conservative with even a moderate number of comparisons. Holm (1979) provided a simple, computationally quick alternative to the Bonferroni method which is computed when `method="holm"` is included in `dunnTest()`.[20]

```
> dunnTest(Wr~gcat,data=ruf90,method="holm")
              Comparison          Z      P.unadj      P.adj
1       stock - quality  -1.6948450 0.090104825 0.36041930
2      stock - preferred  0.9203837 0.357372298 1.00000000
3   quality - preferred   2.7573568 0.005827074 0.03496244
4      stock - memorable  0.6048691 0.545266001 1.00000000
5   quality - memorable   2.5323805 0.011329098 0.05664549
6 preferred - memorable  -0.3717833 0.710054161 0.71005416
```

These results suggest that the median relative weight for quality-length Ruffe is significantly greater than that for preferred-length Ruffe and may be greater than that for memorable-length Ruffe, but the median relative weight does not appear to differ for any other pairs of length categories.

8.4   Further Considerations

Brenden et al. (2003) provided an alternative test, called the *R-test*, for comparing relative weights among groups. The R-test, in contrast to the one-way ANOVA and Kruskal-Wallis tests, is designed to accommodate errors that are not independent and identically distributed. The R-test has not been implemented in R and has not been widely adopted by fisheries scientists. This reticence to use the R-test may be because Pope and Kruse (2007) suggested that parametric tests (e.g., one-way ANOVA), if the residuals are not too nonnormal and homoscedasticity can be shown, or nonparametric tests (e.g., Kruskal-Wallis test) may be adequate for the practicing scientist (but may not be for the research scientist).

The dummy variable regression (also know as analysis of covariance) and related methods discussed in Sections 7.4 and 7.5 can also be used to assess relative condition among groups of fish. In their simplest forms, these methods are used to determine if the weight-length relationship differs among groups, which can then be used to determine if the weight at various lengths differs among groups.

Cade et al. (2011) provided a strong argument against the use of the relative weight index (and, presumably, the other condition metrics) in favor of the use of quantile regression methods. The central premise in their critique is related to the constancy of allometric growth with weight assumed by all standard weight equations (Cone 1989). They argue, and then demonstrate, that this assumption can limit interpretations when condition is compared among groups. They show that quantile regression can provide more insight into differences in condition among groups. An online supplement to their publication shows how to fit quantile regressions with functions in the **quantreg** (Koenker 2015) package.

Aho (2014) noted that the Brunner-Dette-Munk test is similar to the Kruskal-Wallis test but does not assume that the groups have equal variances. This test is computed with `BDM.test()` from **asbio** (Aho 2015a).

Notes

[1]Data manipulations in this chapter require functions from **magrittr** and **dplyr**, which are fully described in Chapter 2.

[2]The constants in Fulton's K are used so that the values are not extremely small decimals.

[3]See Section 7.4 for methods to compare slopes among groups.

[4]As discussed in Section 7.3.1, this back-transformation is subject to bias. However, it is not apparent that the bias-correction factor introduced in that section has been applied to the calculation of the relative condition factor.

[5]Even though the weight-length residuals and the relative condition values relate the same information, Sutton et al. (2000) suggested that the weight-length residuals may have

better statistical properties than the relative condition factor. These properties may be useful when comparing condition among groups of fish (Section 8.3). However, Patterson (1992) and Jakob et al. (1996) criticized the use of weight-length residuals as measures of fish condition.

[6]The use of the common logarithms for the development of the standard weight equations was initially simply a matter of choice. The exponent in the weight-length relationship will be the same regardless of the base of the logarithm, but the intercept will be different.

[7]All species with standard weight equations in `wsVal()` can be seen with `wsVal()` (i.e., with no arguments).

[8]There have been at least three methods for computing standard weight equations, though the most common are the *regression line percentile* (RLP; Murphy et al. 1990) and *empirical percentile* (EmP; Gerow et al. 2005) methods. Details for these methods are in Murphy et al. (1990), Blackwell et al. (2000), and Neumann et al. (2012). Note that there is an ongoing debate in the literature about which of these two methods, if either, can serve as the single method to compute standard weight equations (Gerow et al. 2004, 2005; Rennie and Verdon 2008; Gerow 2010; Ranney et al. 2010; Gerow 2011; Ranney et al. 2011; Neumann et al. 2012).

[9]A maximum length will not be shown if it was not defined in the original source. Similarly, a value for the quadratic coefficient will not be shown unless a quadratic model was used to define the standard weight equation.

[10]The use of `[[]]` here will force the result to be a vector rather than a data.frame (see Section 1.7.1). Vectors are needed when using these values to compute standard weights.

[11]When computing the standard weight, make sure to use the common logarithm of the length variable with the model coefficients and then use the result as the power of 10 (i.e., anti-log).

[12]The reader may find it useful to examine how the different metrics are related in this example. A plot, for example, of the relationship between Wr_i and Kn_i values is constructed with `plot(Wr Kn,data=ruf950,pch=19,col=rgb(0,0,0,1/5))`.

[13]Three examples are Fox (2008), Gotelli and Ellison (2013), and Aho (2014).

[14]The use of Type II sums-of-squares (SS) in `Anova()` and Type I SS in `anova()` was discussed in a note in Section 7.4. These SS and the corresponding tests will not differ in a one-way ANOVA. However, they may differ in a two-way ANOVA (i.e., two factors in the model) if the sample sizes differ among the groups defined by the levels of the two factors). Thus, in this book, there will be no difference in results between Type I and Type II tests. However, following the advice of Fox and Weisberg (2011), Type II tests will be used throughout the book. One should, however, read the discussion of the different types of SS provided by Aho (2014).

[15]The experimentwise error rate is the probability of making a Type I error in *at least* one of the paired comparisons. As the number of paired comparisons increases (i.e., as the number of groups in the one-way ANOVA increases), the experimentwise error rate will increase to rather high levels.

[16]Tukey's HSD method is also implemented in `TukeyHSD()` in base R. The `TukeyHSD()` function is not used here because it requires the use of `aov()` rather than `lm()` and because `glht()` can generalize to other multiple comparison procedures (e.g., Dunnett's method).

[17]Several messages from the output of `summary()` have been removed from this output to save space when printing the book.

[18]The `cld()` function can only be used for Tukey comparisons in `glht()`. The letters from `cld()` are placed beginning at the group with the lowest mean. This order may be reversed by including `decreasing=TRUE` in `cld()`.

[19]The `dunnTest()` in **FSA** is largely a modification of `dunn.test()` from **dunn.test** (Dinno 2015). In addition, several other functions in R packages purport to provide multiple comparison results for the Kruskal-Wallis test. See the help documentation for `dunnTest()` for the names of these other functions.

[20]A discussion of the other methods for controlling the experimentwise error rate is beyond the scope of this introductory book. However, the other methods implemented in `dunnTest()` are described in the help documentation for `p.adjust()`.

9

Abundance from Capture-Recapture Data

A key component of many fisheries assessments is the determination of how many fish are in a population. Fisheries scientists may determine the actual number of fish in the population (i.e., *absolute abundance*) or a metric proportional to the number of fish in the population (i.e., *relative abundance*). Two families of methods commonly used to estimate absolute abundance are capture-recapture and depletion/removal methods. Capture-recapture methods are introduced in this chapter. Depletion/removal methods are introduced in Chapter 10.

Required Packages for This Chapter

Functions used in this chapter require loading the packages shown below.[1]

```
> library(FSA)
> library(dplyr)
> library(Rcapture)
```

9.1 Data Requirements

Fish may be marked such that the fisheries scientist will know whether that fish was previously captured or not. Fish may be given a *batch mark* (e.g., fin-clip or brand) where the individual cannot be specifically identified or an *individual mark* (e.g., anchor, coded-wire, or passive integrated transponder tag) where the individual fish can be specifically identified (Pine et al. 2012). If a batch mark is used, then the numbers of marked (i.e., previously captured) and unmarked fish in a sample are recorded. If an individual mark is used, then the specific tag numbers for fish captured in each sampling event are recorded.

9.1.1 Capture History Format

For individually marked fish, Pollock et al. (1990) recommended that the capture and recapture information be recorded in *capture history* format.[2]

The capture history for an individual fish consists of recording a "1" if the fish was captured in that sampling event or a "0" if it was not, but it was captured in at least one other sampling event (Pine et al. 2003). Thus, the capture history for an individual fish consists of as many "0"s and "1"s as there are sampling events.[3]

For example, the first fish in Table 9.1 was captured (and marked) in the first sample, not captured in the second sample, recaptured in the third sample, and not captured in either the fourth or fifth samples. In contrast, the fifth fish in Table 9.1 was captured (and marked) only in the fifth sample.

TABLE 9.1. Individual capture histories for five fish captured during five sampling events. A "1" means the fish was captured in that event, whereas a "0" means that it was not.

Fish	Sample Event				
	1	2	3	4	5
1	1	0	1	0	0
2	0	1	0	1	0
3	0	1	0	1	0
4	1	0	1	0	0
5	0	0	0	0	1

Capture histories for all fish in a study are entered in either *individual fish* or *frequency* format. In the individual fish format, the capture history for each fish is recorded as one row in the data file (Table 9.1). Thus, the data file in this format has as many rows as uniquely captured fish. In contrast, each row in the frequency format represents a unique capture history with the frequency of fish with that capture history in a final column (Table 9.2). Thus, in the frequency format data file, the number of rows corresponds to the number of unique capture histories and the sum of the frequency column is the total number of fish captured in all sampling events.

TABLE 9.2. Frequency of individuals ("freq") with each unique capture history from five sampling events. This table represents the same five fish shown in Table 9.1.

Sample Event					freq
1	2	3	4	5	
1	0	1	0	0	2
0	1	0	1	0	2
0	0	0	0	1	1

These two capture history formats have been recorded in a variety of ways, largely because there are a number of different softwares available for performing capture-recapture analyses. For example, some programs require the capture history codes for each sample event in a separate column, whereas other programs require the capture history codes as one compact string (e.g., 01010). The methods used in this chapter require capture history information in the individual format with separate columns for each sample event. Other information about the fish (e.g., length, sex, or location) may be included in additional columns.

As an example, Zehfuss et al. (1999)[4] captured Gulf Sturgeon from below the Jim Woodruff Dam on the Apalachicola River (Florida) on 13[5] occasions during the summer of 1985. Fish longer than 45 cm were uniquely tagged. These data, in *Zehfussetal_1985_mod.INP*, were recorded with the capture history codes in "compact form" in one variable, the frequency of fish in a separate variable, and each row ending with a semicolon as required by Program MARK (White et al. 1982).

Care must be taken to properly load these data into R. First, **read.table()** is used instead of **read.csv()** because the text file has columns separated by spaces rather than commas. Second, **header=FALSE** is used because the file does not contain a header row with variable names. Third, the type (or class) of each column is forced to be "character" to avoid dropping the leading zeroes on the capture history strings (e.g., so that "00001" is not interpreted as "1"). Finally, because there was no header, the columns were renamed from the default names with **names()**.

```
> gs.M <- read.table("Zehfussetal_1985_mod.INP",header=FALSE,
                     colClasses=c("character","character"))
> names(gs.M) <- c("ch","freq")
> headtail(gs.M)
               ch freq
1  0000001010001   1;
2  0100000000000   1;
3  0011000000000   1;
74 0000000000001   1;
75 0000000000001   1;
76 0000100101010   1;
```

The **capHistConvert()** function from **FSA** may be used to convert between several common formats of simple capture history data (see the help documentation for **capHistConvert** for a complete list and description of supported formats). This function requires a source data.frame that contains one of the supported formats as the first argument, the format of the source data identified in **in.type=**, and the desired format to be returned in **out.type=**. Furthermore, depending on the format in **in.type=**, the name of the variable that contains the unique fish identifier (e.g., tag number) may be given in `id=` or the name of the variable that contains the frequency of fish may be given in `freq=`. Optionally, one may use `include.id=TRUE` to include unique fish identifiers in a variable in the resultant data.frame (the unique identifiers will be arbitrary numbers when converting from a format where unique fish identifiers do not exist in the source data.frame).

For use with methods described in this chapter, the Gulf Sturgeon data are converted from the "MARK"-type format to the "individual" fish format below. The `var.lbls.pre=` argument is used to set the prefix for the variable labels in the output data.frame (must be a nonnumeric string).

```
> gs.I <- capHistConvert(gs.M,in.type="MARK",freq="freq",
                         out.type="individual",include.id=TRUE,
                         var.lbls.pre="e")
> headtail(gs.I)
   id e1 e2 e3 e4 e5 e6 e7 e8 e9 e10 e11 e12 e13
1   1  1  0  0  0  0  0  0  1  0   1   0   0   0   1
2   2  2  0  1  0  0  0  0  0  0   0   0   0   0
3   3  3  0  0  1  1  0  0  0  0   0   0   0   0
74 74  0  0  0  0  0  0  0  0  0   0   0   0   1
75 75  0  0  0  0  0  0  0  0  0   0   0   0   1
76 76  0  0  0  0  1  0  0  1  0   1   0   1   0
```

9.1.2 Summarizing Capture History Data

Capture history data must be summarized to estimate population parameters with the capture-recapture methods described in Sections 9.2–9.4. Several useful summaries are obtained with `capHistSum()` from **FSA**. The first argument is a data.frame of **ONLY** the capture history data in the individual fish format. The `col2use=` argument offers an efficient method to identify which columns in the data.frame represent only the capture history data. For example, if the capture history data are in columns 2 through 14 in a data.frame with 14 columns (as in `gs.I`) then use either `cols2use=2:14` to select these columns or `cols2use=-1` to *exclude* the first column.[6]

```
> gs.ch <- capHistSum(gs.I,cols2use=-1)
```

The `capHistSum()` object is a list of either two or five objects depending on whether only two or more than two events were recorded in the capture histories. In both cases, the `$caphist` object is a vector that stores the frequency of fish with each unique capture history.

```
> gs.ch$caphist[1:8]  # only first eight shown
```

```
0000000000001 0000000000010 0000000000100 0000000001000
       3             8             2             2
0000000001100 0000000010001 0000000011000 0000000100000
       2             1             1             6
```

Thus, eight fish had the "0000000000010" capture history. The remaining four objects in the `capHistSum()` object — `$sum`, `$methodB.top`, `$methodB.bot`, and `$m.array` — are specific to particular methods and are discussed with those methods in later sections.

9.2 Closed Population, Single Recapture

9.2.1 Single Group

The most simple and common capture-recapture study occurs when M fish are collected from a closed population, marked with either a batch or individual mark, and returned to the population. A subsequent sample of n fish is collected and the number of previously seen or marked fish (m) is determined. Under strict assumptions (Section 9.2.3), the ratio of M to the total population size (N) is equal to the ratio of m to n. This equality of ratios is algebraically solved for N to produce the well known *Petersen* equation for estimating population size.[7] The simple Petersen estimate is biased for small samples. Chapman (1951), however, showed that

$$\widehat{N} = \frac{(M+1)(n+1)}{(m+1)} - 1 \tag{9.1}$$

is an unbiased estimator of N when $(M+n) \geq N$, or is nearly unbiased when $m > 7$ (Krebs 1999).

In most simple situations, confidence intervals for N can be constructed from a hypergeometric distribution. Historically, however, confidence intervals for N have been approximated from other distributions depending on characteristics of the data. Seber (2002) suggested that if more than 10% of fish in the second sample are recaptured fish (i.e., $\frac{m}{n} > 0.10$), then a binomial distribution should be used. Otherwise, if $m < 50$, then a Poisson distribution should be used, or if $m > 50$, then a normal distribution should be used.[8]

The Petersen and related modifications for estimating N are conducted with `mrClosed()` from **FSA**.[9] This function requires M, n, and m as the first three arguments (in that order). The default estimator is the original Petersen estimate, but the Chapman modification is used by including `method="Chapman"`.[10] The $\widehat{N}$ is extracted from the `mrClosed()` object with `summary()`. An estimated standard error (SE) for $\widehat{N}$ is provided if `incl.SE=TRUE` is used in `summary()`.

A confidence interval (CI) for N is extracted from the `mrClosed()` object with `confint()`. By default, the distribution used to construct the CI follows Seber's guidelines described previously. However, a specific distribution may be selected by including one of `"hypergeometric"`, `"binomial"`, `"Poisson"`, or `"normal"` in `type=`. Which distribution was used to calculate the CI is not shown unless `verbose=TRUE` is included in `confint()`.

As an example, Wisconsin Department of Natural Resources biologists marked 2555 Walleye greater than 304 mm from Sand Lake on one night. They returned the next night and captured 274 Walleye greater than 304 mm, with 92 of those fish being recaptured from the previous night's effort.

```
> sl1 <- mrClosed(2555,274,92,method="Chapman")
> summary(sl1,incl.SE=TRUE)
        N     SE
[1,] 7557 622.5
> confint(sl1,verbose=TRUE)

The binomial distribution was used.
     95% LCI 95% UCI
[1,]    6454    8973
```

Thus, the estimated number of Sand Lake Walleye longer than 304 mm is 7557 (95% CI: 6454, 8973).

If capture histories, rather than summarized counts, are recorded for two events, then they should be summarized with `capHistSum()` as shown in Section 9.1.2. The resulting object is then submitted as the first argument to `mrClosed()` which will automatically extract the values for M, n, and m. This functionality is illustrated below using just the fifth and sixth sample events for Gulf Sturgeon as an example. The code below also demonstrates how the `summary()` and `confint()` results can be column-bound together with `cbind()` to form a more succinct display.

```
> gs.ch56 <- capHistSum(gs.I,cols2use=5:6)
> gs.mr56 <- mrClosed(gs.ch56,method="Chapman")
> cbind(summary(gs.mr56,incl.SE=TRUE),confint(gs.mr56))
       N   SE 95% LCI 95% UCI
[1,] 74 33.9      22     145
```

9.2.2   Separate Groups

Fish may be separated into groups to control for different catchabilities among groups (Pine et al. 2003) or to provide more detailed results. To perform this analysis, the M, n, and m values for each group are entered into separate vectors, which are then the first three arguments to `mrClosed()`. A vector of labels for each group may be given to `labels=`. In addition to providing an abundance estimate for each group, the separate estimates are combined to provide an estimate of total abundance.[11]

As an example, Sand Lake Walleye longer than 304 mm were separated into those fish shorter and longer than 381 mm.

```
> group <- c("305-380","381+")
> M <- c(1929,626)
> n <- c( 222, 52)
> m <- c(  77, 15)
> sl2 <- mrClosed(M,n,m,labels=group,method="Chapman")
> cbind(summary(sl2,incl.SE=TRUE),confint(sl2,incl.all=TRUE))
           N    SE 95% LCI 95% UCI
305-380 5517 490.4    4656    6649
381+    2076 415.5    1439    3211
All     7593 642.8    6333    8853
```

Thus, the estimated number of Sand Lake Walleye longer than 304 mm is 7593 (95% CI: 6333, 8853).

9.2.3 Assumptions

Valid application of the Petersen and related modifications depends on five assumptions being reasonably met (Seber 2002; Hayes et al. 2007; Pine et al. 2012):

1. The population is physically (i.e., no immigration or emigration) and demographically (i.e., no recruitment or mortality) closed.
2. Marks or tags are neither lost nor missed.
3. Marked fish returned to the population mix randomly with unmarked fish.
4. All fish within a sample have an equal probability of capture.
5. Fish behavior or vulnerability does not change after being marked.

Fisheries personnel attempt to meet the first assumption by maintaining a very short period between the first and second samples. Seber (2002) outlined a chi-square test to detect recruitment between the two samples (requires length categories to have been recorded). Tag loss may be assessed by double-tagging fish with the observed proportional tag loss used to adjust the final abundance estimate (Pine et al. 2003). Equal probability of capture is often difficult to ensure because of the selectivity of most fishing gears. This assumption may be grossly met by using several different gears to capture fish or stratifying by length (or other grouping) as shown above (Pine et al. 2003). Differential capture probability over time, due to behavioral changes in the fish, or among fish, can generally only be assessed (and modeled) with more than one recapture event and using more sophisticated models (see Section 9.3.2).

9.3 Closed Population, Multiple Recaptures

9.3.1 Schnabel and Schumacher-Eschmeyer Methods

The single recapture study of Section 9.2 can be extended to multiple capture and recapture events if the unmarked fish in a sample are marked (with either a batch or an individual mark) and returned to the population along with the previously marked fish. For sample event i, the numbers of captured fish (n_i), previously marked fish (m_i), and marked fish returned to the population (R_i) are recorded.[12]

The number of marked fish extant in the population just prior to taking sample i (M_i) must also be known. Because the population is assumed to be closed, M_i is the sum of the number of **newly** marked fish returned to the population on all samples prior to sample i (i.e., $M_i = \sum_{j=1}^{i-1}(R_j - m_j)$, for $i \geq 2$ and $M_1 = 0$).

Two methods for estimating abundance from these data are the *Schnabel* (Schnabel 1938) and *Schumacher-Eschmeyer* (Schumacher and Eschmeyer 1943) methods (but see Section 9.3.2). The Schnabel estimator, as modified by Chapman (1951) to reduce bias, is

$$\widehat{N} = \frac{\sum_{i=1}^{k} n_i M_i}{\left(\sum_{i=1}^{k} m_i\right) + 1} \tag{9.2}$$

which is essentially a weighted average of Petersen estimates at each recapture event. The Schumacher-Eschmeyer estimator is

$$\widehat{N} = \frac{\sum_{i=1}^{k} n_i M_i^2}{\sum_{i=1}^{k} m_i M_i} \tag{9.3}$$

and is based on minimizing the weighted sum-of-squares between the proportion of marked animals in a sample (i.e., $\frac{m_i}{n_i}$) and the unknown proportion of marked animals in the population. Confidence intervals for N are constructed from a t distribution for the Schumacher-Eschmeyer method and the Schnabel method if $\sum m_i \geq 50$, or from the Poisson distribution otherwise (Seber 2002).[13]

An estimate of N, and associated confidence interval, using the Schnabel or Schumacher-Eschmeyer methods is obtained with mrClosed().[14] The n_i,

m_i, and either the M_i or R_i must be given to `mrClosed()` in one of two ways. First, if these values have already been summarized (as would likely be the case if a batch mark was used), then individual vectors of these values are given to `n=`, `m=`, and either `M=` or `R=`. Alternatively, if the data are recorded as capture histories, then the object saved from `capHistSum()` may be the first argument to `mrClosed()` and the n_i, m_i, and M_i values will be extracted. The Schnabel method is chosen with `method="Schnabel"`[15], whereas the Schumacher-Eschmeyer method is selected with `method="SchumacherEschmeyer"`. The $\widehat{N}$ is extracted from the `mrClosed()` object with `summary()` and the confidence interval, computed following the recommendations of Seber (2002), is extracted with `confint()`.

The capture histories for Gulf Sturgeon were summarized in Section 9.1.2. The `$sum` object in the `capHistSum()` object contains the n_i, m_i, R_i, and M_i values (this object also contains three other summary values that will be explained in Section 9.3.2).

```
> headtail(gs.ch$sum)  # only first & last 3 rows (events) shown
    n  m  R  M  u  v  f
1   4  0  4  0  4  0 44
2   9  1  9  4  9  4 21
3   4  1  4 12  3  2  7
11  6  4  6 63  2  5  0
12 11  3 11 65  8 11  0
13 12  9  0 73  3 12  0
```

These values are automatically used when the `capHistSum()` object is the first argument to `mrClosed()`.

```
> gs.mr1 <- mrClosed(gs.ch,method="Schnabel")
> cbind(summary(gs.mr1),confint(gs.mr1,verbose=TRUE))
The Poisson distribution was used.
       N 95% LCI 95% UCI
[1,] 103      78     139
```

Thus, using the Schnabel method, the estimated number of Apalachicola River Gulf Sturgeon longer than 45 cm is 103 (95% CI: 78, 139).

9.3.2 Log-Linear Model Methods

Some researchers have argued that the assumption (required for the methods in Section 9.3.1) that all animals have the same capture probability within each sampling event is rarely met (Otis et al. 1978; Pine et al. 2003). To address this issue, Otis et al. (1978) described a hierarchical set of models that allow for four types of capture probabilities.[16]

- M_0 is the *null model*. It represents no heterogeneity in capture probability (i.e., all fish are always equally likely to be caught). Abundance (N) and the capture probability for an individual fish (p) can be estimated.
- M_t is the *time-varying* model. The capture probability may vary from one sampling event to another, but it is the same for all fish within any sampling event. This is the same assumption for the Petersen and related modifications, Schnabel, and Schumacher-Eschmeyer methods. Abundance and the probability of capture for fish *within* a sampling event (p_i) can be estimated.
- M_b is the *behavior-varying* model. The capture probability may vary following a previous capture. For example, the probability of capture may be lower ("trap-shy") or higher ("trap-happy") for fish that have been previously captured. Abundance and the probabilities of capture for the first (p) and all subsequent captures (c) can be estimated.
- M_h is the *individual-varying* model. In this model, individual fish have different capture probabilities, but those probabilities are constant for all sampling events. In theory, this model has parameters for population abundance and capture probabilities of *every* fish. As this would result in more parameters than observations, a simplifying distribution for the probabilities of capture is used (Hayes et al. 2007), with a variety of distributions having been proposed. Thus, abundance and an overall measure ("average") of the probability of capture for individual fish can be estimated.

The remaining models described by Otis et al. (1978) combine the three types of heterogeneous capture probabilities. For example, the M_{th} model allows the probability of capture to vary across sampling events and among individual fish. These combination models are generally more difficult to fit and require higher probabilities of capture (Pine et al. 2012). The M_{tb} model is nonlinear and cannot be fit with the same algorithms as the other models (Baillargeon and Rivest 2007). Finally, an estimate of population abundance cannot be obtained from the M_{tbh} model (Hayes et al. 2007).

Hayes et al. (2007) illustrated a brute-force maximum likelihood approach to estimate parameters for some of these models. However, log-linear (e.g., Poisson regression) methods have been used to maximize the likelihoods for the models and to estimate parameters and related SE (e.g., Baillargeon and Rivest 2007). This modeling framework has been implemented in **Rcapture** (Rivest and Baillargeon 2014)[17] and is demonstrated below with the Gulf Sturgeon data.

Baillargeon and Rivest (2007) argued that exploratory methods should be used to informally deduce, prior to model fitting and selection, the type of heterogeneity, if any, present in the capture history data. They described two plots to aid this exploration. The first plot is of $log\left(\frac{f_j}{\binom{k}{j}}\right)$ against j, where f_j is the number of fish captured j times and k is the total number of sampling events. The second plot is of $log(u_i)$ against i, where u_i is the number of fish captured for the first time in sample i. Baillargeon and Rivest

TABLE 9.3. Form of the log(scaled f_j) versus j and $\log(u_i)$ versus i plots for different capture-recapture models. Note that L means "linear," ~L means "almost linear," C means "concave upward" or "convex," and none means that the plot has no definitive form. Adapted from Baillargeon and Rivest (2007).

	Model					
Plot	M_0	M_t	M_b	M_h	M_{th}	M_{bh}
log(scaled f_j) vs j	L	~L	none	C	~L / C	none
$\log(u_i)$ vs i	L	none	L	C	none	C

(2007) provided descriptors for the shapes of these plots in the face of different types of heterogeneity (Table 9.3). Both plots (Figure 9.1) are constructed by submitting the `capHistSum()` object to `plot()`.

```
> plot(gs.ch)
```

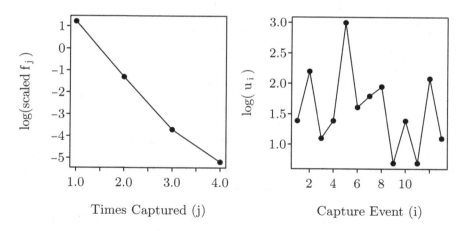

FIGURE 9.1. Plots for diagnosing heterogeneity in capture probabilities for the Apalachicola River Gulf Sturgeon.

The plot of $log(\text{scaled} f_i)$ values is either approximately linear or very slightly concave and the plot of $log(u_i)$ values has no definitive shape (Figure 9.1). Thus, according to Table 9.3, the capture probabilities of the Apalachicola River Gulf Sturgeon are likely heterogeneous across sampling events (i.e., M_t) and, possibly, by individual (i.e., M_{th}).

The log-linear analogues to the models of Otis et al. (1978) are fit with `closedp.t()` from **Rcapture**. The first argument is a data.frame of **ONLY** capture histories, either in the individual or frequency format.[18] Recall that the first column of `gs.I` contains a unique fish identifier that is not part of the capture history data and, thus, must be excluded (i.e., with `[,-1]`) for use here.

```
> gs.cp <- closedp.t(gs.I[,-1])   # be patient
```

The model fit results and estimates of abundance and related SE are extracted by appending `$results` to the `closedp.t()` object. Note that there are several versions of the M_h and M_{th} models depending on the distribution used to model the individual heterogeneity.

```
> gs.cp$results
```

	abundance	stderr	deviance	df	AIC	BIC
M0	110.54	10.60	201.97	8189	295.26	299.92
Mt	108.60	10.17	166.76	8177	284.04	316.67
Mh Chao (LB)	118.55	17.13	201.16	8185	302.44	316.43
Mh Poisson2	110.84	10.70	201.95	8188	297.23	304.23
Mh Darroch	121.78	23.03	201.53	8188	296.82	303.81
Mh Gamma3.5	132.08	38.31	201.40	8188	296.68	303.67
Mth Chao (LB)	117.12	16.60	165.72	8172	293.00	337.28
Mth Poisson2	108.90	10.26	166.73	8176	286.01	320.97
Mth Darroch	121.46	22.86	166.15	8176	285.43	320.39
Mth Gamma3.5	133.23	39.16	165.99	8176	285.27	320.23
Mb	182.09	103.28	199.16	8188	294.44	301.44
Mbh	135.68	51.86	196.61	8187	293.89	303.21

Simple[19] information criteria (IC) metrics combine a measure of the *lack-of-fit* of a model to data with a penalty for the number of parameters in that model. A measure of lack-of-fit is equal to two times the inverse of the likelihood function for the data and model, evaluated at the parameters that maximize that likelihood function. In other words, the parameters that maximize the likelihood function will minimize two times the inverse of that function. The penalty for number of parameters in the model (k) varies between various IC metrics. For example, Akaike's IC (AIC) uses $2k$ as the penalty, whereas the Bayesian IC (BIC) uses $k log_e(n)$. Smaller IC values indicate a better model (i.e., low lack-of-fit and number of parameters). The BIC metric tends to favor models with fewer parameters than the AIC metric.[20]

The model with the lowest IC value is the model, from the suite of models considered, that best fits the data (Burnham and Anderson 2002). If AIC is used in this case (see `AIC` column in the output from `gs.cp$results` above), M_t is the best model. However, several of the M_{th} models have similar AIC values, which suggests that there may be, in addition to the time-varying heterogeneity, a heterogeneity in capture probabilities at the individual level.

A confidence interval for N is computed via the profile likelihood method[21] within `closedpCI.t()`. The first argument is the data.frame of capture histories exactly as described for `closedp.t()`. The model from which to construct the confidence interval is entered into `m=` and may be one of `"M0"`, `"Mt"`, `"Mh"`, or `"Mth"`.

```
> ( gs.ci <- closedpCI.t(gs.I[,-1],m="Mt") )
Number of captured units: 76

Poisson estimation and model fit:
   abundance stderr deviance   df    AIC     BIC infoFit
Mt     108.6   10.2  166.759 8177 284.04 316.671      OK


Multinomial estimation, 95% profile likelihood conf. interval:
   abundance infCL supCL infoCI
Mt     107.5  91.5 132.1     OK
```

Thus, the estimated number of Apalachicola River Gulf Sturgeon longer than 45 cm is 107 (95% CI: 92, 132).[22] A visual (not shown) of the profile likelihood is produced by submitting the `closedpCI.t()` object to `plotCI()`.[23]

Estimates of the other parameters from *ALL* models are extracted by appending `$parameters` to the `closedp.t()` object. The parameters for a particular model are retrieved by also appending the name of the model. For example, the parameters for M_0 and M_t are retrieved with the following code.

```
> gs.cp$parameters$M0
             N          p
estimate: 110.5416 0.08559255
> gs.cp$parameters$Mt
             N         p1         p2         p3         p4
estimate: 108.5956 0.0368339 0.08287628 0.0368339 0.04604238
             p5         p6         p7         p8         p9
estimate: 0.2210034 0.1197102 0.05525085 0.1012932 0.09208475
            p10        p11        p12        p13
estimate: 0.0736678 0.05525085 0.1012932 0.1105017
```

Thus, under the M_t model, the estimated capture probability for the first sampling event is 0.0368, whereas the estimated capture probability for the sixth event is 0.1197.

9.3.3 Assumptions

The Schnabel and Schumacher-Eschmeyer methods have the same assumptions as the single recapture methods of Section 9.2.3. The log-linear model methods also share the same assumptions, except that the probability of capture does not have to be constant within or among sampling events.

With multiple recapture events, possible violations of the assumptions are more easily detected. One diagnostic tool for assessing the assumptions is a plot of $\frac{m_i}{n_i}$ versus M_i. If this plot is not linear, then one or more of the assumptions has been violated. Unfortunately, the shape of the plot does not indicate which assumption has been violated. This plot is constructed from the `mrClosed()` object with `plot()`. A loess smoother is added to the plot with `loess=TRUE`, but this is only useful with several (≥ 8) recapture events. Figure 9.2 shows a generally linear but widely varying relationship between the proportion of recaptures in the sample and the number marked in the population, which suggests one of the model assumptions has been violated for the Gulf Sturgeon data (which, from the work in Section 9.3.2, is a time-varying heterogeneity in capture probabilities).

```
> plot(gs.mr1,loess=TRUE)
```

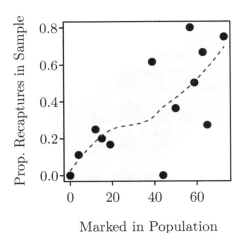

FIGURE 9.2. Proportion of recaptured fish at each sampling event versus number of extant marks in the population prior to each sampling event for Apalachicola River Gulf Sturgeon.

The probabilities of capture in each sampling event as estimated with M_t can also be plotted against i to better understand capture probability heterogeneity (Figure 9.3). This plot is constructed below and illustrates the strong among sample heterogeneity in capture probabilities of Gulf Sturgeon that was identified in Section 9.3.2.

```
> plot(1:12,gs.cp$parameters$Mt[2:13],type="l",ylim=c(0,0.24),
        xlab="Sampling event",ylab="Probability of Capture")
> abline(h=gs.cp$parameters$M0[,"p"],lty=2)
```

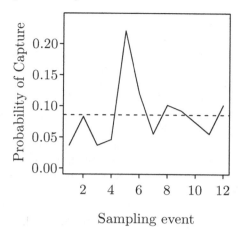

FIGURE 9.3. Probability of capture by sampling event for the Apalachicola River Gulf Sturgeon as estimated from the M_t model. The dashed horizontal line is the probability of capture estimated from the M_0 model.

9.4 Open Populations

9.4.1 Jolly-Seber Method

Methods to estimate abundance described in Sections 9.2 and 9.3 are only applicable to populations that are closed to recruitment, mortality, immigration, and emigration. The assumption of a closed population is usually maintained by having samples with little time between them (i.e., not enough time for substantial recruitment, mortality, immigration, or emigration to occur). However, many realistic situations occur with samples that span periods of time where the assumption of a closed population is not reasonably assured. The method described in Jolly (1965), modified in Seber (1965), and collectively referred to as the Jolly-Seber method is used to estimate abundance at the time of sample event i (N_i) when three or more samples of marked fish have been collected from an open population.[24]

The sampling scheme for a Jolly-Seber study is similar to that described in Section 9.3.1 (i.e., multiple sampling events are used and previously unmarked fish are marked and returned to the population). However, fish are usually individually marked because the Jolly-Seber calculations are based on knowing when (which sampling event) a fish was last caught and whether it was caught again in subsequent sampling events.[25]

The capture history data are summarized in a *Method B table* (Leslie and Chitty 1951), which is referred to here as having two parts. The *top of the Method B table* is a contingency table of the frequency of marked fish in sample i that were last caught in previous sample j (m_{ji}). For example, $m_{38} = 2$ shows that two fish captured in the eight sample were last caught in the third sample. All cells for which $j \geq i$ are "missing" (i.e., NA in R) because those observations are physically impossible.

The *bottom of the Method B table* contains the numbers of recaptured marked fish (m_i), unmarked fish (u_i), total fish (n_i), and total marked fish returned to the population (R_i) for each of the i sampling events.

Derivations of maximum likelihood estimates for the number of marked fish extant in the population just prior to sample i (i.e., M_i) and N_i, and their standard errors, are not trivial. These derivations are, however, well-detailed in Pollock et al. (1990) and Krebs (1999) and are not discussed in detail here. Note, though, that estimates of N_1 and N_k, where k is the number of sampling events, cannot be estimated.

As an example, Harding et al. (2010) captured Cutthroat Trout from Auke Lake (Alaska) during the summers of 1998–2006. All Cutthroat Trout greater than 180 mm were given a PIT tag. Sampling occurred on multiple days within each year, but each year was considered as a single sampling event. Individual fish capture histories, along with a unique fish identification code in the first column, are recorded in *CutthroatAL.csv*.

```
> cutty <- read.csv("CutthroatAL.csv")
> headtail(cutty,n=2)
       id y1998 y1999 y2000 y2001 y2002 y2003 y2004 y2005 y2006
1       1     0     0     0     0     0     0     0     0     1
2       2     0     0     0     0     0     0     0     0     1
1683 1683     1     1     1     0     0     0     0     0     0
1684 1684     1     1     1     0     0     0     0     0     0
```

The Cutthroat Trout capture histories are summarized with capHistSum() as described in Section 9.1.2. The "top" and "bottom" of the Method B table are extracted by appending $methodB.top and $methodB.bot, respectively, to the capHistSum() object.

```
> cut.ch <- capHistSum(cutty,cols2use=-1)
> cut.ch$methodB.top
    i=1 i=2 i=3 i=4 i=5 i=6 i=7 i=8 i=9
j=1  NA  22   4   0   0   0   0   0   0
j=2  NA  NA  90   4   2   0   0   0   0
j=3  NA  NA  NA  37  13   1   0   0   0
j=4  NA  NA  NA  NA  43   2   1   0   0
j=5  NA  NA  NA  NA  NA  96   2   2   0
j=6  NA  NA  NA  NA  NA  NA  88  10   1
j=7  NA  NA  NA  NA  NA  NA  NA  40   4
j=8  NA  NA  NA  NA  NA  NA  NA  NA  13
j=9  NA  NA  NA  NA  NA  NA  NA  NA  NA
```

```
> cut.ch$methodB.bot
   i=1 i=2 i=3 i=4 i=5 i=6 i=7 i=8 i=9
m    0  22  94  41  58  99  91  52  18
u   89 330 198 192 201 271 199  82 122
n   89 352 292 233 259 370 290 134 140
R   89 352 292 233 259 370 290 134   0
```

The calculations for the Jolly-Seber method are performed by submitting the capHistSum() object to mrOpen() from **FSA**.[26] The $\widehat{M}_i$ and $\widehat{N}_i$, and their standard errors, are extracted from the mrOpen() object with summary() using parm=c("M","N"). Intermediate values used to compute these estimates are extracted if verbose=TRUE is included (not shown).

```
> cut.mr <- mrOpen(cut.ch)
> summary(cut.mr,parm=c("M","N"))
         M M.se     N  N.se
i=1    NA   NA    NA    NA
i=2  36.6  6.4 561.1 117.9
i=3 127.8 13.4 394.2  44.2
i=4 120.7 20.8 672.2 138.8
i=5  68.3  4.1 301.0  21.8
i=6 117.5  7.3 436.1  30.3
i=7 175.1 24.6 553.7  84.3
i=8 100.2 24.7 255.3  65.4
i=9    NA   NA    NA    NA
```

Confidence intervals for the N_i, computed using asymptotic large-sample theory (i.e., assuming a normal distribution; Jolly 1965), are extracted from the mrOpen() object with confint() using parm="N"[27].

```
> confint(cut.mr,parm="N")
     N.lci N.uci
i=1    NA    NA
i=2 330.0 792.1
i=3 307.6 480.8
i=4 400.2 944.2
i=5 258.3 343.6
i=6 376.7 495.5
i=7 388.4 719.1
i=8 127.2 383.4
i=9    NA    NA
```

Thus, for 2003 (i.e., $i = 6$), the estimated abundance of Auke Lake Cutthroat Trout greater than 180 mm was 436 (95% CI: 377, 496).

9.4.2   Log-Linear Model or Cormack-Jolly-Seber Method

Cormack (1989) described a log-linear model approach to estimate N_i and the probability of capture in sample i (i.e., p_i) from multiple samples of individually marked fish from an open population, possibly with variable capture probabilities.[28] This approach considers three models related to fish behavior following capture:

- *No "trap-effect" model.* Fish have an equal capture probability before and after having been captured.
- *Homogeneous "trap-effect" model.* The capture probability of fish differs before and after initial capture but does not differ across multiple samples before, or multiple samples after, initial capture.
- *Full "trap-effect" model.* The capture probability of fish differs depending on the number of times captured.

The specifics of these models and how parameters are estimated are beyond the scope of this book. However, these models have been implemented in openp() of **Rcapture**. A simple example of the use of openp() is demonstrated below for Auke Lake Cutthroat Trout.[29]

In its simplest form, openp() requires a data.frame with **ONLY** the capture history information as its first argument.[30] Again, note the use of [,-1] to remove the first column with fish identification codes and leave just the capture history information. The deviance (two times the negative log likelihood), residual degrees-of-freedom, and AIC for the no trap-effect model are extracted by appending $model.fit to the openp() object.

```
> cut.op <- openp(cutty[,-1])
> cut.op$model.fit
                 deviance     df       AIC
fitted model  86.33589     487   315.8965
```

Similarly, the same statistics for the two trap effect models are extracted by appending $trap.fit to the openp() object.

```
> cut.op$trap.fit
                                         deviance     df       AIC
model with homogenous trap effect  85.19608     486   316.7567
model with trap effect              81.12745     481   322.6881
```

The lower AIC value for the no trap-effect model suggests that there is not a substantial "trap effect" evident in these data.

The $\widehat{N_i}$ derived from the no trap-effect model are extracted by appending $N to the openp() object.

```
> cut.op$N
          estimate      stderr
period 1        NA          NA
period 2 586.6667 127.65695
period 3 398.7134  45.36168
period 4 693.5642 144.08699
period 5 305.2628  23.03419
period 6 439.8398  31.27062
period 7 563.0519  86.38136
period 8 266.8107  69.48155
period 9        NA          NA
```

Similarly, the $\hat{p}_i$ are extracted by appending $capture.prob to the object.

```
> cut.op$capture.prob
           estimate       stderr
period 1         NA           NA
period 2 0.6000000 0.13211565
period 3 0.7323556 0.08621986
period 4 0.3359458 0.07205980
period 5 0.8484494 0.06723064
period 6 0.8412153 0.06229378
period 7 0.5150502 0.08177597
period 8 0.5022288 0.13432230
period 9         NA           NA
```

Approximate confidence intervals for the model parameters (e.g., N_i) are obtained by assuming asymptotic normality for the distributions of the estimates. In this code, qnorm() finds the value on a standard normal distribution that has the given proportional area lower. The with() function is used so that N can be used rather than cut.op$N.

```
> conf.level <- 0.95
> z <- qnorm(.5+conf.level/2)
> with(cut.op,cbind(N.LCI=N[,"estimate"]-z*N[,"stderr"],
                    N.UCI=N[,"estimate"]+z*N[,"stderr"]))
            N.LCI     N.UCI
period 1       NA        NA
period 2 336.4636 836.8697
period 3 309.8061 487.6206
period 4 411.1588 975.9695
period 5 260.1166 350.4089
period 6 378.5505 501.1291
period 7 393.7476 732.3563
period 8 130.6293 402.9920
period 9       NA        NA
```

Thus, for 2003 (i.e., `period` 6), the estimated abundance of Auke Lake Cutthroat Trout greater than 180 mm was 440 (95% CI: 379, 501) and the estimated probability of capture was 0.841 (95% CI: 0.719, 0.963; results not shown).

9.4.3 Assumptions

The Jolly-Seber method assumes the following (Krebs 1999; Seber 2002):

1. Marks or tags are neither lost nor missed.
2. All fish within a sample have an equal probability of capture.
3. Every marked individual has the same probability of surviving from sample i to sample $(i + 1)$. If survival rate estimates are to apply to all, rather than just marked, individuals, then it is also assumed that the probability of survival is the same for both marked and unmarked fish.
4. Sampling time is negligible in relation to intervals between samples.

Krebs (1999) outlined several tests for the critical assumption of equal capture probability for marked and unmarked fish. These methods have not yet been implemented in R. However, the log-linear models approach can be used to detect specific types of unequal capture probabilities.

A form of residual plot (Figure 9.4) from the log-linear models is created by submitting the `openp()` object to `plot()` (shown below with some modifications for the main title and the type and color of points plotted). In this case, one large residual for an individual captured three times is apparent, as is a distinct narrowing of the range of residuals for fish captured six or more times. Baillargeon and Rivest (2007) demonstrated how to explore what effect removing individuals has on the model fit and parameter estimates.

```
> plot(cut.op,main="",pch=19,col=rgb(0,0,0,1/6))
```

9.5 Further Considerations

Estimating fish abundance is an important endeavor and a great deal of energy and ingenuity has been expended to develop estimates that are accurate and precise. The methods developed in this chapter are foundational and have been greatly extended in the fisheries, wildlife, and ecological literature. These extensions are generally more complex than those shown here. Some of these methods have been collated into R packages.

RMark (Laake 2013) is primarily an interface to Program MARK, which many consider the "gold standard" software for complex capture-recapture models. **RMark** uses the formula and design matrix attributes of R to ease

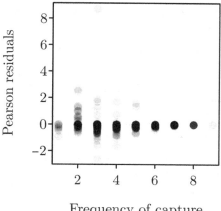

FIGURE 9.4. Residual plot from the log-linear model fit to the Auke Lake Cutthroat Trout data.

development of complex capture-recapture models that are then exported and fit in Program MARK. **RMark** requires an understanding of Program MARK which is extensively documented (over 1000 pages!) at http://www.phidot.org/software/mark/.

Other packages that perform some sort of capture-recapture analysis are **marked** (Laake et al. 2013), **unmarked** (Fiske and Chandler 2011), **secr** (Efford 2015), **mra** (McDonald 2015), and **mrds** (Laake et al. 2015).

Notes

[1]Data manipulations in this chapter require functions from **magrittr** and **dplyr**, which are fully described in Chapter 2.

[2]Some authors and programs (e.g., Program MARK; White et al. 1982) use *encounter history* rather than *capture history*. Capture history is used here.

[3]The capture history information described here is the simplest possible form, where fish are simply recorded as having been captured or not. More complex forms, for example, using "-1" to recognize mortality of a tagged fish, are utilized with other software. Consult the documentation for Program MARK (White et al. 1982) for more information on these more complex structures.

[4]These data were also used in Pine et al. (2012).

[5]Zehfuss et al. (1999) actually recorded data on 16 sampling occasions. Unfortunately, methods from **Rcapture** did not work with all 16 sampling events. Thus, three sampling events where only a single fish was captured (first, fourth, and seventh events in the original data) were combined with adjacent sampling events to reduce the number of sampling events to 13. These modified data are used throughout this chapter.

[6]Note that columns that do NOT contain the capture history can also be given in cols2ignore=. For example, use cols2ignore=1 to not use the first column. The ad-

vantage to using `cols2ignore=` is that it allows a vector of column **names**. For example, `col2ignore="id"` will result in the `id` not being used.

[7]The motivation for and derivations of the equations for the Petersen and related methods are fully described in many other sources (e.g., Ricker 1975; Krebs 1999; Borchers et al. 2004; Hayes et al. 2007; Pine et al. 2012).

[8]The use of these distributions is meticulously detailed in Seber (2002).

[9]Similar functionality for computing Petersen and related estimates of N is available in `mrN.single()` from **fishmethods**.

[10]Bailey (1951, 1952) also modified the Petersen estimate for situations where it is possible that the recaptured fish may be counted more than once (e.g., if marks are observed remotely without handling the fish). Ricker (1975) provided a slight modification to the Chapman (1951) modification by ignoring the final subtraction of one individual in Equation (9.1). The Bailey or Ricker modification is used by including `method="Bailey"` or `method="Ricker"`, respectively, to `mrClosed()`.

[11]The estimate of total abundance is simply the sum of the abundance estimates for all groups. The variance for the sum of multiple values is the sum of the variances plus two times the covariances between pairs of values. Newman and Martin (1983) showed that the covariance between pairs of mark-recapture estimates of N by group is negligible and, thus, the variance of the sum is approximately the sum of the variances. Thus, the standard error for the estimate of total abundance is approximately equal to the square root of the sum of the squared standard errors (i.e., the variances) for all groups.

[12]If the capture-recapture data are recorded as capture histories, the methods herein assume that all fish were returned with a mark; that is, $R_i = n_i$. This does not cause a loss of generality in these methods because, under the assumption of a closed population, the number of fish not returned to the population can be tallied separately and added to the abundance estimate to provide an estimate of the initial population size.

[13]The motivations for and derivations of these methods are provided in the original papers and in several other common resources (e.g., Ricker 1975; Krebs 1999; Hayes et al. 2007).

[14]Similar functionality for computing Schnabel and Schumacher-Eschmeyer abundance estimates is available in `schnabel()` from **fishmethods**.

[15]By default, the Chapman (1951) modification of the Schnabel method is used. The original equation of Schnabel (1938) is used by including `chapman.mod=FALSE`.

[16]The work of Otis et al. (1978) has been greatly expanded. The interested reader will develop a deeper understanding by consulting other more detailed resources (e.g., White et al. 1982; Pollock et al. 1990; White and Burnham 1999; Williams et al. 2002; Borchers et al. 2004; Chao and Huggins 2005).

[17]A variety of softwares other than R have been developed to fit the Otis et al. (1978) models. Program MARK (White and Burnham 1999), and its associated Program CAPTURE, is the most cited software for this purpose in the fisheries literature. See Section 9.5 for more information about the related **RMark** package.

[18]The use of the frequency format is not demonstrated here. However, if the data are in frequency format, then `closedp.t()` requires that the frequencies be in the last column and that `dfreq=TRUE` be included.

[19]Information criteria are used extensively in the ecological literature. This introduction to IC metrics greatly simplifies their foundation and use in model selection. The interested reader should enter the literature on this topic with Burnham and Anderson (2002).

[20]There is considerable confusion and debate about whether to use the AIC or BIC metric. The interested reader could begin an exploration of this topic with Burnham and Anderson (2004) or Aho et al. (2014).

[21]The profile likelihood method for constructing confidence intervals is described generally in Hilborn and Mangel (1997), Bolker (2008), and Haddon (2011).

[22]Louis-Paul Rivest, author of **Rcapture**, suggests using $\widehat{N}$ from the multinomial distribution in most cases (personal communication). Further note that an estimate of N that is a composite computed from all fitted models may be computed by summing the estimates of N multiplied by Akaike weights from each model. More information about *model*

averaging and Akaike weights is found in Burnham and Anderson (2002). Akaike weights are computed with `ICtab()` from **bbmle** (Bolker and R Development Core Team 2014).

[23] Note that several packages have functions called `plotCI()`. If those other packages are loaded after **Rcapture**, then simply typing `plotCI()` may attempt to use the "wrong" `plotCI()`. The **Rcapture** version is forced to be used with `Rcapture::plotCI()`.

[24] The Jolly-Seber method can also be used to estimate apparent survival from event i to $i+1$ (ϕ_i) and apparent recruitment from event i to $i+1$ (B_i). Discussion and examples for computing ϕ_i are in Section 11.3.1.

[25] The Jolly-Seber method may be used with fish that are marked with a capture-event-specific mark. However, this may be cumbersome with many sampling events. Thus, individual marks are usually used.

[26] It is also possible to perform the Jolly-Seber calculations by manually entering (from previously summarized results) the top and bottom portions of the Method B table into separate matrices and then supplying those matrices as the first two arguments to `mrOpen()`. See the help documentation for `mrOpen` for an example of this functionality.

[27] Manly (1984) criticized the default confidence intervals suggested by Jolly (1965) and provided an alternative method. The method of Manly (1984) appears arbitrary and has not been widely adopted. However, Manly's method is computed using `type="Manly"` in `mrOpen()`.

[28] The log-linear model method of Cormack (1989) can also be used to estimate apparent survival from event i to $i+1$ (ϕ_i) and apparent recruitment from event i to $i+1$ (B_i). Estimates of ϕ_i are illustrated in Section 11.3.2.

[29] More detailed examples of the use of `openp()` are provided by Baillargeon and Rivest (2007).

[30] As with `closedp()`, if the capture history information is in frequency rather than individual format, then the frequency information must be in the last column and `dfreq=TRUE` is included in `openp()`.

10

Abundance from Depletion Data

Two families of methods are commonly used to estimate absolute abundance – capture-recapture and depletion/removal methods. Capture-recapture methods were described in Chapter 9. Depletion/removal methods are described here.

Required Packages for This Chapter

Functions used in this chapter require loading the packages shown below.[1]

```
> library(FSA)
> library(dplyr)
```

10.1 Leslie and DeLury Methods

The relationship between catch or catch per unit effort (CPE) and either cumulative catch or cumulative effort can be exploited to estimate the initial abundance (N_0) for populations that experience enough fishing that the catch of fish declines with successive catches.

The *Leslie Method* uses

$$\frac{C_i}{f_i} = qN_0 - qK_{i-1} \tag{10.1}$$

where C_i is the catch for sample i, f_i is the fishing effort for sample i, q is the constant *catchability coefficient* (i.e., the fraction of the population that is removed by one unit of fishing effort), and K_{i-1} is the cumulative catch prior to sample i.[2] The left-hand side of Equation (10.1) is shown here as CPE, but it can be catch (Hilborn and Walters 2001).[3]

Equation (10.1) is a linear model where $\frac{C_i}{f_i}$ is the response variable, K_{i-1} is the explanatory variable, $-q$ is the slope, and qN_0 is the intercept (Figure 10.1). Thus, the *negative of the slope* is an estimate of the catchability coefficient ($\widehat{q}$). The estimated initial population size ($\widehat{N}_0$) is the intercept divided by $\widehat{q}$. Visually, $\widehat{N}_0$ is the point where the linear model intercepts the *x-axis*, or the total cumulative catch such that the CPE would equal zero (Figure 10.1).

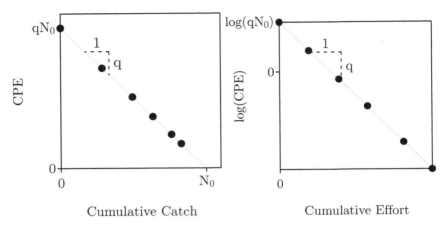

FIGURE 10.1. Idealized plot of the decline in catch per unit effort (CPE) with increasing cumulative catch (Left; Leslie Model) and decline in log CPE with increasing cumulative effort (Right, DeLury Model). Visual representations of the catchability coefficient, q, and initial population size, N_0, are shown.

If $q < 0.02$, then q behaves as an instantaneous rate and the model developed by DeLury (1947) is more appropriate than Equation (10.1) (Ricker 1975). The DeLury model is

$$log\left(\frac{C_i}{f_i}\right) = log(qN_0) - qE_{i-1} \qquad (10.2)$$

where E_{i-1} is the cumulative fishing effort prior to sample i.[4] This model is again in the form of a linear model (Figure 10.1) where $log\left(\frac{C_i}{f_i}\right)$ is the response variable, E_{i-1} is the explanatory variable, $-q$ is the slope, and $log(qN_0)$ is the intercept. Thus, the *negative of the slope* is again $\widehat{q}$ and $\widehat{N}_0$ is the anti-logged intercept divided by $\widehat{q}$.

The confidence interval for q is the same, except for sign change, as the confidence interval for the slope of the linear model. The confidence interval for N_0 is not straightforward, as $\widehat{N}_0$ is the ratio of two statistics. However, Seber (2002) provided equations for calculating the standard error of $\widehat{N}_0$ for both the Leslie and DeLury methods. Confidence intervals for q and N_0 are computed using standard large-sample methods (i.e., using the t distribution) with the standard errors.

The Leslie method is illustrated here with removals of age-0 Largemouth Bass from stands of Eurasian Watermilfoil in Lake Guntersville (Alabama; Maceina et al. 1993). Fish were collected from one 0.11 hectare enclosure with six consecutive samples of 10 minutes of electrofishing each, with the exception that the fifth sample was only 6 minutes due to equipment difficulties. The number of fish removed in the six samples was 7, 7, 4, 1, 2, and 1.

The catch and effort data were entered into a data.frame and variables of CPE, K_{i-1}, natural log of CPE, and E_i were created. The K_{i-1} and E_i were computed from the catch and effort vectors, respectively, with pcumsum() from **FSA**.[5]

```
> mac <- data.frame(catch=c(7,7,4,1,2,1),
                    effort=c(10,10,10,10,6,10)) %>%
    mutate(cpe=catch/effort,K=pcumsum(catch),
           logcpe=log(cpe),E=pcumsum(effort))
> mac
  catch effort       cpe  K    logcpe  E
1     7     10 0.7000000  0 -0.3566749  0
2     7     10 0.7000000  7 -0.3566749 10
3     4     10 0.4000000 14 -0.9162907 20
4     1     10 0.1000000 18 -2.3025851 30
5     2      6 0.3333333 19 -1.0986123 40
6     1     10 0.1000000 21 -2.3025851 46
```

Following the methods described in Section 7.3 for fitting linear regression models, Equation (10.1) is fit with lm() and the slope and intercept are extracted (and saved to an object) with coef(). The $\widehat{q}$ and $\widehat{N}_0$ are then computed from the saved slope and intercept.[6]

```
> lm1 <- lm(cpe~K,data=mac)
> ( cf1 <- coef(lm1) )
(Intercept)           K
 0.78643157 -0.03019312
> ( q.hat1 <- -cf1[["K"]] )
[1] 0.03019312
> ( N0.hat1 <- cf1[["(Intercept)"]]/q.hat1 )
[1] 26.04672
```

Estimates from the DeLury method are computed similarly, except that vectors of cumulative effort and log CPE are used (note that the intercept is raised to the power of e with exp()).

```
> lm2 <- lm(logcpe~E,data=mac)
> ( cf2 <- coef(lm2) )
(Intercept)           E
-0.25118318 -0.03990633
> ( q.hat2 <- -cf2[["E"]] )
[1] 0.03990633
> ( N0.hat2 <- exp(cf2[["(Intercept)"]])/q.hat2 )
[1] 19.49264
```

Thus, the Leslie and DeLury methods, respectively, estimate that there are 26 and 19 age-0 Largemouth Bass in this enclosure.

The Leslie and DeLury methods are more efficiently computed using depletion() from **FSA**.[7] The first two arguments are the vectors of catch and corresponding effort (in that order). The Leslie method is the default, but the DeLury method may be selected with method="DeLury".

```
> d2 <- depletion(mac$catch,mac$effort)
```

The estimates of q and N_0 are extracted from the depletion() object with summary() or coef() and confidence intervals are constructed with confint().

```
> cbind(summary(d2),confint(d2))
     Estimate Std. Err.     95% LCI    95% UCI
No 26.0467186 3.3733507 16.6807956 35.412642
q   0.0301931 0.0068508  0.0111722  0.049214
```

Thus, there appears to be 26 (95% CI: 17, 35) age-0 Largemouth Bass in this enclosure with a catchability coefficient of 0.030 (95% CI: 0.011, 0.049).

A plot of CPE versus cumulative catch (or log(CPE) versus cumulative effort if the DeLury method was used) with the best-fit line and parameter estimates superimposed (Figure 10.2) is constructed with plot().

```
> plot(d2)
```

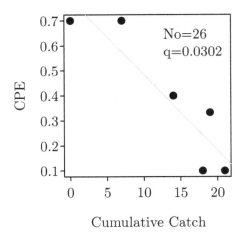

FIGURE 10.2. Plot of CPE versus the prior cumulative catch for age-0 Largemouth Bass in a Lake Gunterville enclosure.

10.1.1 Assumptions

The Leslie and DeLury methods are based on the following assumptions related to the fish and fishery:

1. The population is physically and demographically closed.
2. Catchability is constant over the period of removals.
3. Enough fish are removed to substantially reduce the CPE.
4. All fish are equally vulnerable to the method of capture (common sources of error include gear saturation and trap-happy or trap-shy individuals).
5. The units of effort are independent (i.e., individual capture units such as nets or traps do not compete with each other).

The Leslie method also assumes that the catches remove more than 2% of the population. Finally, the usual assumptions of simple linear regression (Section 7.3.3) also apply for both models.

The fitted line plot for the Largemouth Bass data (Figure 10.2) appears to have a slight curvature which suggests that these data may not meet the assumptions of the Leslie model. It is, however, difficult to definitively make this conclusion given the small sample size.

10.2 K-Pass Removal Methods

10.2.1 Equal Catchability Among Passes

In the *k-pass removal* method, population abundance is estimated for a closed population by sampling the population k times *with the same amount of effort*. On each sampling "pass," the number of individuals captured (C_i) is recorded and the individuals are physically removed from the population. If the probability of capture (p) is constant for all fish across all samples, then an estimate of the initial abundance is the smallest $N_0 \geq T$ that solves

$$\frac{N_0 + 1}{N_0 - T + 1} \prod_{i=1}^{k} \frac{kN_0 - X - T + \beta + k - i}{kN_0 - X + \alpha + \beta + k - i} \leq 1 \qquad (10.3)$$

where $T = \sum_{i=1}^{k} C_i$ is the total number of individuals captured in all removal events, $X = \sum_{i=1}^{k} (k-i)C_i$, and α and β are parameters from a beta distribution that forms a prior distribution for the capture probability (Carle and Strub 1978).[8] Once $\widehat{N}_0$ is found by iteratively solving Equation (10.3), then

$$\widehat{p} = \frac{T}{k\widehat{N}_0 - X} \tag{10.4}$$

Zippin (1956, 1958) provided equations for the standard errors of N_0 and p.

Other removal methods have been proposed by Zippin (1956, 1958), Seber (2002) for two and three passes, and Robson and Regier (1968) for two passes. However, Seber (2002) noted that the Carle and Strub (1978) method has a smaller bias and variance, is more robust to assumption violations, and will not mathematically "fail."[9] Additionally, the Carle and Strub (1978) method generally provides more accurate estimates of N_0 and p than the other methods (Hedger et al. 2013). While these other methods can be computed using the functions described below, only the Carle and Strub (1978) method is illustrated.

As an example, scientists with the Bonneville Power Administration conducted a three-pass removal for Rainbow Trout in various stretches of the West Fork of the Hood River (Oregon) on September 21, 1994 as part of an environmental impact statement prior to constructing the Pelton Ladder on the Hood River (Olsen et al. 1995). Several sections of tributary streams were blocked off by 3 mm mesh netting (effectively closing the population) and three electrofishing passes were made through each stretch. The number of Rainbow Trout in two size groups was recorded for each section. In the Lake Branch section, 187, 77, and 35 Rainbow Trout less than 85 mm were removed in the three passes. The three catches are entered into a vector below.

```
> ct <- c(187,77,35)
```

The iterations required to solve Equation (10.3) for N_0 are conducted with `removal()` from **FSA**. The first argument is a vector that contains the catches from each removal pass. The Carle-Strub method is used by default, but other methods may be selected with `method=` (see the help documentation for `removal` for other supported methods). When `method="CarleStrub"`, values for α and β are included in `alpha=` and `beta=`, respectively (both default to 1 which represents a noninformative uniform prior distribution). Values of $\widehat{N}_0$ and $\widehat{p}$ are extracted with `summary()` and corresponding confidence intervals are constructed with `confint()`.

```
> pr1 <- removal(ct)
> cbind(summary(pr1),confint(pr1))
      Estimate Std. Error    95% LCI    95% UCI
No 323.0000000  8.4158244 306.5052874 339.494713
p    0.5772201  0.0355731   0.5074982   0.646942
```

Thus, there appears to be 323 (95% CI: 307, 339) Rainbow Trout less than 85 mm in this stretch.

Use of `removal()` as described above is tedious if the fisheries scientist needs to estimate abundance from several sets of removal data (e.g., multiple sites or sizes of fish). As a small example, the catches of Brook Trout from three passes at each of three stations (MU10, MU13, and MU27) of "reach 3" of Reader Creek (Utah) in 2004 (Birchell 2007) were entered into the data.frame below.

```
> (df <- data.frame(sta=c("MU10","MU13","MU27"),
                    p1=c(19,75,20),p2=c(14,19,11),p3=c(9,5,3)) )
    sta p1 p2 p3
1 MU10 19 14  9
2 MU13 75 19  5
3 MU27 20 11  3
```

The `apply()` function is used to apply a function to each row or column of a matrix or data.frame as described in Section 1.9.2. In this case, `just.ests=TRUE` is included as the last argument to `apply()` to force `removal()` to return just the parameter estimates, SEs, and confidence interval values.

```
> res <- apply(df[,-1],MARGIN=1,FUN=removal,just.ests=TRUE)
```

Unfortunately, `apply()` returns values in columns rather than rows. Thus, the result from `apply()` must be transposed with `t()`. Finally, the results are coerced to a data.frame with the station names from the original data.frame appended as the first column.

```
> ( res <- data.frame(sta=df$sta,t(res)) )
    sta  No     No.se    No.LCI    No.UCI          p       p.se
1 MU10   55 11.727984  32.01357   77.98643 0.3716814 0.12613965
2 MU13  100  1.470384  97.11810 102.88190 0.7557252 0.04549002
3 MU27   36  2.507203  31.08597  40.91403 0.5964912 0.10295277
      p.LCI      p.UCI
1 0.1244522 0.6189106
2 0.6665664 0.8448840
3 0.3947075 0.7982749
```

Thus, at station MU10 there appears to be 55 (95% CI: 32, 78) Brook Trout.

10.2.2   Catchability Differs in First Pass

A common problem with the methods described in the previous section is that the probability of capture is not often constant among all passes. The most common violation of this assumption is that the probability of capture on the first pass (p_1) differs from the probability of capture on all subsequent passes (p).

Schnute (1983) described two models for removal data and a likelihood ratio test to determine which model best represents the data. The first model that Schnute described was from Moran (1951), where p was constant across all removal events. He then described a new model where p_1 differed from p. Both models described by Schnute derive $\widehat{N}_0$ by minimizing, with respect to N_0, the negative log-likelihood function

$$N_0 log(N_0) - T log(T) - (N_0 - T) log(N_0 - \widehat{T}) - log\binom{N_0}{T} \qquad (10.5)$$

where $\widehat{T}$ is the predicted total catch.

The two models differ in how predicted catches are computed. The predicted catches for the first model are computed as $\widehat{C}_i = N_0 \widehat{p}(1 - \widehat{p})^{i-1}$ where the predicted probability of capture is from Equation (10.4). For the second model, the predicted catch for the first removal pass is simply C_1 and the predicted catches for the subsequent passes are $\widehat{C}_i = N_0 \widehat{p}(1 - \widehat{p}_1)(1 - \widehat{p})^{i-2}$ for $i = 2, \ldots, k$ and where $\widehat{p}_1 = \frac{C_1}{N_0}$ and

$$\widehat{p} = \frac{T - C_1}{(k-1)(N_0 - C_1) - (X - (k-1)C_1)} \qquad (10.6)$$

Once $\widehat{N}_0$ is found, $\widehat{p}$ is found with Equation (10.4) for the Moran model or Equation (10.6) for the Schnute model. The $\widehat{p}_1$ for the Schnute model is estimated with $\frac{C_1}{\widehat{N}_0}$.

Importantly, Schnute (1983) also showed that two times the difference in minimum negative log-likelihood values (i.e., plug $\widehat{N}_0$ into Equation (10.5)) between these two models follows a chi-square distribution with one degree-of-freedom. This provides a statistical test to determine the most appropriate model for the removal data. A small p-value favors the Schnute model and the conclusion that the probability of capture for the first sample is different than the constant probability of capture for all subsequent samples.

The Moran and Schnute models are fit by including `method="Moran"` and `method="Schnute"`, respectively, in `removal()`. The minimum negative log-likelihood value is extracted by appending `$min.nlogLH` to a `removal()` object.[10] The p-value for Schnute's likelihood ratio test is computed by submitting two times the difference in negative log-likelihood values between the two models to `pchisq()` along with `df=1` (degrees-of-freedom) and `lower.tail=FALSE` (to compute the upper tail probability).

```
> M1 <- removal(ct,method="Moran")
> S1 <- removal(ct,method="Schnute")
> ( teststat <- 2*(M1$min.nlogLH-S1$min.nlogLH) )
[1] 0.01939134
> pchisq(teststat,df=1,lower.tail=FALSE)
[1] 0.8892504
```

In this case, there is very little evidence ($p = 0.8893$) for a different probability of capture on the first sample. Thus, one of the constant probability of capture models (e.g., the Carle-Strub model) is likely adequate.

Notes

[1] Data manipulations in this chapter require functions from **dplyr**, which are fully described in Chapter 2.

[2] Ricker (1975) modified Equation (10.1) by replacing K_{i-1} with K_i, which is $K_{i-1} + \frac{C_i}{2}$ (i.e., adding half of the catch from the current sample). This modification is used in depletion() by including `Ricker.mod=TRUE`.

[3] Hilborn and Walters (2001) also noted that the left-hand side of Equation (10.1) may be collected independently from the data used to measure cumulative catch on the right-hand side. For example, CPE could be measured from acoustic surveys, whereas the cumulative catch could come from a trapnet fishery.

[4] Again, Ricker (1975) modified Equation (10.2) by replacing E_{i-1} with E_i, which is $E_{i-1} + \frac{f_i}{2}$ (i.e., adding half of the effort from the current sample). This modification is used in depletion() by including `Ricker.mod=TRUE`.

[5] The pcumsum() function computes the *prior cumulative sum* or the cumulative sum prior to the current position in the vector. The cumsum() function in base R computes the cumulative sum prior to and including the current position in the vector.

[6] Note that the use of [[]] in this code is an R "trick" to remove names from the values. Without doing this, the result saved in q.hat would be annoyingly named as (intercept). See Section 1.7.1.

[7] The deplet() function from **fishmethods** may also be used to compute the Leslie and DeLury estimates.

[8] Most applications of the Carle and Strub (1978) method have assumed no prior information about the distribution of the capture probability and, thus, have used a uniform distribution defined by $\alpha = \beta = 1$. However, for example, Hedger et al. (2013) used informative priors derived from a "calibration site" that improved parameter accuracy when used with an open population.

[9] Some of the other methods can mathematically "fail" if, for example, the number of fish removed on the last pass is greater than or equal to the number of fish removed on the first pass.

[10] Parameter estimates, but not standard errors, from the Moran and Schnute models are extracted with summary(). Profile likelihood confidence intervals, as described by Schnute (1983), are extracted with confint().

11

Mortality Rates

Mortality rates are a measure of the rate at which fish disappear from a population. A primary goal for management of exploited populations is to regulate harvest such that the total mortality rate is below that which allows the population to persist. Thus, estimates of mortality rates are critical information for fish managers.

Methods for estimating the total mortality rate of fish form the bulk of this chapter. Later sections provide methods for separating total mortality into components due to fishing and nonfishing.

Required Packages for This Chapter

Functions used in this chapter require loading the packages shown below.[1]

```
> library(FSA)
> library(car)        # Before dplyr to reduce conflicts with MASS
> library(dplyr)
> library(magrittr)
> library(Rcapture)
```

11.1 Total Mortality Definitions

Total annual mortality rate (A) is the proportion of fish in a population closed to immigration, recruitment, and emigration that die in one year. Total annual survival rate (S) is the proportion of the same population that survive during the year. Thus, $A + S = 1$. If the catch of fish is proportional to the size of the population, then A is computed with

$$A = \frac{C_t - C_{t+1}}{C_t} = 1 - \frac{C_{t+1}}{C_t} \tag{11.1}$$

where C_t is the catch of fish at time t.

Population models may be expressed in a continuous form such that the rate of change parameter is an *instantaneous rate* (Haddon 2011). In mortality models, the *instantaneous total mortality rate* (Z) is a measure of how the number of individuals declines in an imperceptibly short period of time (i.e., in an "instant"). Alternatively, Z also relates how the natural log number of individuals declines annually, which gives rise to the formula

$$Z = log(C_t) - log(C_{t+1}) \qquad (11.2)$$

Instantaneous rates are mathematically convenient because the exponential nature of most mortality models becomes linear when expressed on the log scale. Additionally, Z can be partitioned into components of mortality that are additive (see Section 11.4). This convenience comes at the cost of Z being difficult to interpret (Haddon 2011). For example, what does it mean that the log number of individuals declines by 0.693? Fortunately, Z is easily converted to A with

$$A = 1 - e^{-Z}$$

Thus, the largely uninterpretable $Z = 0.693$ corresponds to 50.0% ($A = 1 - e^{-0.693}$) of the population dying on an annual basis.[2]

11.2   Total Mortality from Catch Curve Data

11.2.1   Data Requirements

As shown in Equations 11.1 and 11.2, total mortality rates can be computed from catches of fish that are separated by one year. However, more synthetic and precise estimates of mortality rates can be computed from catches of fish at several pairs of successive ages.[3] Methods that use the frequencies of fish of each age to estimate mortality rates are commonly referred to as *catch curve methods*.

Catch-at-age data are derived from two common sources. First, *longitudinal* data may be collected by following the same cohort of fish through time. Second, fish may be collected at one point in time, coming from many cohorts, to generate *cross-sectional* data. Vectors of cross-sectional and longitudinal catch-at-age data are equal if the mortality rate is constant across time and among ages, recruitment is the same for each cohort, and the probability of capture is constant across time and among ages (Table 11.1).

The typical exponential decline in numbers of fish at each age is linear when the natural logarithm of catch is plotted against age (Figure 11.1). These plots, called *catch curves*, commonly have three regions of interest. The *ascending left limb* represents ages of fish that are not yet fully vulnerable, the *peak* represents the age where the fish have just become fully vulnerable, and the *descending right limb* begins at the peak and represents ages of fish that are fully vulnerable to the gear used in the fishery. Mortality rates are

TABLE 11.1. The hypothetical catch of fish by age and capture year. The longitudinal catch of the 2008 year-class is shown by the diagonal cells highlighted in darker gray. The cross-sectional catch in the 2015 capture year is shown by the column of cells highlighted in lighter gray. All data were simulated with constant recruitment, mortality rates, and probabilities of capture.

	Capture Year							
Age	2008	2009	2010	2011	2012	2013	2014	2015
0	500	500	500	500	500	500	500	500
1	350	350	350	350	350	350	350	350
2	245	245	245	245	245	245	245	245
3	171	171	171	171	171	171	171	171
4	120	120	120	120	120	120	120	120
5	84	84	84	84	84	84	84	84
6	59	59	59	59	59	59	59	59

estimated from the ages on the descending limb of the catch curve, as these fish are a representative sample of those ages in the population.

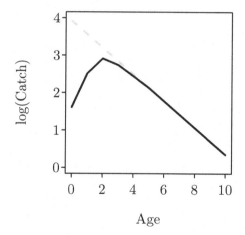

FIGURE 11.1. Idealized plot of the natural log of catch versus age illustrating the typical ascending (younger ages) and descending (older ages) limbs of a catch-curve. The dashed gray line represents the idealized log catch if all age-classes had been fully vulnerable to the fishing gear.

A variety of rules have been proposed to objectively define the descending limb of a catch curve (Smith et al. 2012). Most commonly, the peak has been defined as the age, or one year older than the age, with the greatest catch.[4] Several rules, reviewed by Smith et al. (2012), have also been used to truncate the right side of the descending limb in an attempt to reduce the bias in estimates of Z associated with few fish of older ages. Which rules to use depends on the method of analysis used (some guidance is provided below, but see Smith et al. (2012) for a thorough discussion of the rules).

The catch curve methods will be demonstrated in this section with the catches-at-age of Brook Trout captured in Tobin Harbor (Michigan) from 1996–1998 (Quinlan 1999). These data are simple and, thus, were entered directly into a data.frame in R.[5] A plot of log catch versus age (Figure 11.2) suggests that ages 2 through 6 define the descending limb of the catch curve for these data.

```
> bkt <- data.frame(age=0:6,ct=c(39,93,112,45,58,12,8))
> plot(log(ct)~age,data=bkt,
        xlab="Age (yrs)",ylab="Log Catch",pch=19)
```

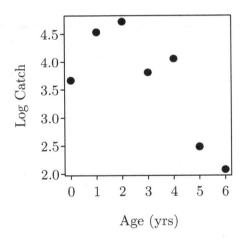

FIGURE 11.2. Scatterplot of log catch versus estimated age for Tobin Harbor Brook Trout.

11.2.2 Chapman-Robson Method

Chapman and Robson (1960) (and Robson and Chapman (1961)) provided a method to estimate S from catch-at-age data. Application of their method requires recoding the ages such that the first fully recruited age on the descending limb of the catch curve is set to 0 (Table 11.2).

The Chapman-Robson estimate of the annual survival rate[6] is

$$\widehat{S} = \frac{T}{n+T-1} = \frac{\bar{T}}{1+\bar{T}-\frac{1}{n}}$$

TABLE 11.2. Cross-sectional catch-at-age of Tobin Harbor Brook Trout, including recoded ages for the Chapman-Robson method. Note that the descending limb of the catch curve for these data begins at age-2.

Actual Age	Recoded Age	Catch
0	–	39
1	–	93
2	0	112
3	1	45
4	2	58
5	3	12
6	4	8

where n is the total number of fish observed on the descending limb of the catch curve, T is the sum of the recoded ages of fish on the descending limb of the catch curve (i.e., the sum of catch multiplied by recoded age), and $\bar{T}$ is the mean recoded age of fish on the descending limb of the catch curve (i.e., $\bar{T} = \frac{T}{n}$) (Miranda and Bettoli 2007). The standard error of this estimate was provided by Miranda and Bettoli (2007) as

$$SE_{\widehat{S}} = \sqrt{\widehat{S}\left(\widehat{S} - \frac{T-1}{n+T-2}\right)}$$

Hoenig et al. (1983) showed that an unbiased estimate of Z could be obtained from $\widehat{S}$ with

$$\widehat{Z} = -log(\widehat{S}) - \frac{(n-1)(n-2)}{n(T+1)(N+T-1)}$$

Hoenig et al. (1983) also provided an estimate for $SE_{\widehat{Z}}$, but Smith et al. (2012, p. 960) modified this by adjusting for a variance inflation factor (c)

$$SE_{\widehat{Z}} = \frac{1-e^{-\widehat{Z}}}{\sqrt{ne^{-\widehat{Z}}}}\sqrt{c}$$

where c is the "usual chi-square goodness-of-fit test statistic divided by the square root of the df." Approximate confidence intervals (CI) are computed from large-sample theory (i.e., using a normal distribution and the SE).

The Chapman-Robson method is implemented in `chapmanRobson()` from **FSA** using three arguments.[7] The first argument is a formula of the form `catch~age`, where `catch` and `age` generically represent the variables that contain the catch and age data. The data.frame that contains `catch` and `age` is given in `data=`. This data.frame does NOT have to be restricted to only ages on the descending limb of the catch curve. However, if ages are included in the data.frame that are not on the descending limb, then the ages on the descending limb must be given as a vector in ages2use=.[8]

```
> thcr <- chapmanRobson(ct~age,data=bkt,ages2use=2:6)
```

The $\widehat{S}$, $SE_{\widehat{S}}$ and bias-corrected estimates of $\widehat{Z}$ and $SE_{\widehat{Z}}$ are extracted from the chapmanRobson() object with summary().[9] Corresponding confidence intervals for both S and Z are extracted with confint().[10] These results are column-bound together below with cbind() for succinctness.

```
> cbind(summary(thcr),confint(thcr))
    Estimate Std. Error     95% LCI     95% UCI
S 49.4600432  2.3260749 44.9010202 54.0190662
Z  0.7018264  0.1153428  0.4757586  0.9278941
```

Finally, a plot that illustrates the Chapman-Robson calculation (Figure 11.3) is obtained by submitting the chapmanRobson() object to plot().

```
> plot(thcr)
```

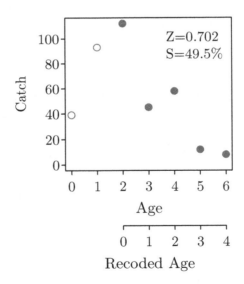

FIGURE 11.3. Catches at estimated ages for Tobin Harbor Brook Trout. The solid points were used to compute the Chapman-Robson estimates of S and Z.

11.2.3   Regression Methods

11.2.3.1   Single Estimate

The descending limb of the catch-curve is linear under a series of assumptions (Section 11.2.4). The negative slope of a linear regression model fit to the natural logarithm of catch or catch per unit effort at each age on the descending limb is an estimate of Z (Miranda and Bettoli 2007). Thus, an estimate of Z is derived by filtering the data.frame to include only those ages on the descending limb of the catch-curve, fitting a simple linear regression (see Section 7.3) of the natural logarithm of catches on ages with `lm()`, and extracting the negative of the estimated slope with `coef()`.

```
> tmp <- filter(bkt,age>=2) %>% mutate(lnct=log(ct))
> lm1 <- lm(lnct~age,data=tmp)
> coef(lm1)
(Intercept)        age
   6.069939   -0.659987
```

From these results, $\widehat{Z}$ for the Tobin Harbor Brook Trout is 0.660.

Maceina and Bettoli (1998) suggested that a weighted regression should be used to reduce the relative impact of older ages with fewer fish on $\widehat{Z}$. As weights, they used the natural log number of fish predicted at each age with the unweighted regression fit to the descending limb of the catch curve. These weights are constructed using `predict()` with the `lm()` object from the unweighted regression. After being appended to the data.frame, these values are used as weights for the regression using `weights=` in `lm()`.

```
> tmp %<>% mutate(wts=predict(lm1))
> lm2 <- lm(lnct~age,data=tmp,weights=wts)
> coef(lm2)
(Intercept)        age
  6.0085938   -0.6430183
```

Thus, using the weighted regression, $\widehat{Z}$ for the Tobin Harbor Brook Trout is 0.643.

Confidence intervals for Z are computed using usual linear regression theory for the slope.

```
> confint(lm2)
                2.5 %      97.5 %
(Intercept)  4.266116    7.751072
age         -1.094109   -0.191928
```

Thus, the 95% confidence interval for Z is from 0.192 to 1.094.

For convenience, these methods are implemented in `catchCurve()` from

FSA,[11] using the same three arguments as used in chapmanRobson(). The weighted regression is fit if weighted=TRUE is given to catchCurve(). The $\widehat{Z}$, $\widehat{A}$, $SE_{\widehat{Z}}$, and the t test statistic and p-value (i.e., Pr(>|t|)) for the test of whether Z equals zero or not are extracted with summary(). Corresponding confidence intervals are obtained with confint().

```
> thcc <- catchCurve(ct~age,data=bkt,ages2use=2:6,weighted=TRUE)
> cbind(summary(thcc),confint(thcc))
     Estimate Std. Error t value   Pr(>|t|)   95% LCI    95% UCI
Z   0.6430183  0.1417433  4.5365 0.02004993   0.191928   1.094109
A  47.4296703         NA      NA         NA  17.463369  66.516206
```

Finally, a plot that illustrates the catch curve and the best-fit line is constructed with plot(). Note the use of pos.est= to move the estimates label from the default top-right corner.

```
> plot(thcc,pos.est="bottomleft")
```

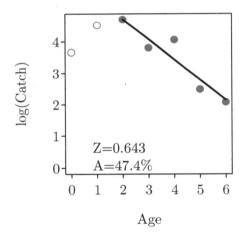

FIGURE 11.4. Natural log of catch at estimated ages for Tobin Harbor Brook Trout including a best-fit line fit to ages 2–6.

11.2.3.2 Comparing Estimates

Given that $\widehat{Z}$ comes from the negative slope on the descending limb of a catch curve, dummy variable regression (DVR; Fox 1997) or analysis of covariance (ANCOVA; Pope and Kruse 2007) can be used to determine if Z differs significantly between two or more groups of fish. Fitting a DVR was introduced and extensively described in Section 7.4.

As an example, Barada (2009) provided the catch-at-age of Channel Catfish from two sections (Central and Lower) of the Platte River (Nebraska).
These data are loaded below, the natural log of catch is appended to the
data.frame, and the data.frame is reduced to only those fish that were age-3
and older (i.e., those fish that were fully recruited to the gear). The DVR is
fit and the resulting ANOVA table is obtained as described in Section 7.4.

```
> d <- read.csv("CCatfishNB.csv") %>% mutate(lnct=log(catch))
> d3 <- filter(d,age>=3)
> lmCC <- lm(lnct~age*loc,data=d3)
> Anova(lmCC)
Anova Table (Type II tests)

Response: lnct
          Sum Sq Df F value    Pr(>F)
age       32.274  1 225.119  7.61e-11
loc        3.699  1  25.804 0.0001115
age:loc    3.026  1  21.105 0.0002995
Residuals  2.294 16
```

The interaction (age:loc) p-value ($p = 0.0003$) provides strong evidence for
a difference in Z (i.e., slope) between the two sections.

As discussed in Section 7.4, the coefficient for the covariate (age) is an
estimate of the slope for the first group as defined in the factor variable (loc).
The coefficient for the interaction (age:locLower) is an estimate of the **difference** in slopes between the two sections. Thus, the sum of the covariate
and the interaction coefficients is the slope for the second group (i.e., lower
section).[12] These relationships, along with recalling that $\widehat{Z}$ is the negative of
the slope, allow one to compute $\widehat{Z}$ for Channel Catfish from both sections of
the river.[13]

```
> (tmp <- coef(lmCC) )
 (Intercept)           age     locLower age:locLower
   4.6101556    -0.2901575    2.7689124   -0.2470670
> -1*tmp[["age"]]                          # Z for Central sect
[1] 0.2901575
> -1*(tmp[["age"]]+tmp[["age:locLower"]]) # Z for Lower sect
[1] 0.5372245
```

A visual of regression fits for both locations (Figure 11.5) is constructed as
follows (following directions from Section 7.4).

```
> clrs <- c("black","gray70")
> plot(lnct~age,data=d,pch=19,col=clrs[loc],
       xlab="Age",ylab="log(Catch)")
```

```
> axis(1,12)          # to put 12 on the x-axis
> xs <- 3:13
> ysc <- predict(lmCC,data.frame(age=xs,loc="Central"))
> lines(ysc~xs,lwd=2,col=clrs[1])
> ysl <- predict(lmCC,data.frame(age=xs,loc="Lower"))
> lines(ysl~xs,lwd=2,col=clrs[2])
> legend("topright",levels(d3$loc),pch=19,col=clrs,
         cex=0.9,bty="n")
```

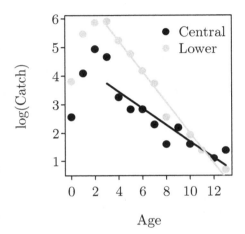

FIGURE 11.5. Catch curves for Channel Catfish from the Central and Lower sections of the Platte River. Regressions were fit to age-3 and older fish.

11.2.4 Catch Curve Assumptions

The catch curve methods make the following assumptions:

1. The population is closed to emigration and immigration.
2. Recruitment is constant among years (if cross-sectional data are used or if estimates from longitudinal data will be extended to other year-classes).
3. Z is constant across ages and years for ages on the descending limb of the catch curve.
4. Vulnerability (if catch data are used) and catchability (if CPE data are used) of the fish is constant across ages and years for ages on the descending limb of the catch curve.
5. The sample is not biased regarding any specific age group(s).

In addition, the usual linear regression assumptions (see Section 7.3.3) must be met for the regression method.

Assumption violations may be detected by careful examination of the catch curve for nonlinearities (and, perhaps, with a residual plot as described in Section 7.3.3). For example, nonconstant recruitment or mortality may be indicated by either a change in slope or wide variability along the descending limb of the catch curve.

11.3   Total Mortality from Capture-Recapture Data

Estimates of *apparent survival* from one time period to the next can be estimated for open populations using capture-recapture data (Section 9.4). These methods estimate *apparent survival* because they cannot, without other information (e.g., telemetry data), distinguish between mortality and emigration (Pine et al. 2003). Traditionally, these methods used ϕ rather than S to represent the survival rate, a convention that will also be used here.

The sampling scheme and resulting data for these methods was described in Section 9.4. However, be reminded that these methods generally require data from individually tagged fish recorded in the *capture history format* (Section 9.1.1). The same capture history data for Cutthroat Trout greater than 180 mm captured from Auke Lake (Alaska) during the summers of 1998–2006 (Harding et al. 2010) used in Section 9.4 are used here. These data are loaded, capture histories are summarized, and the Method B table observed as shown below (this code was thoroughly explained in Section 9.4).

```
> cutty <- read.csv("CutthroatAL.csv")
> headtail(cutty)
       id y1998 y1999 y2000 y2001 y2002 y2003 y2004 y2005 y2006
1       1     0     0     0     0     0     0     0     0     1
2       2     0     0     0     0     0     0     0     0     1
3       3     0     0     0     0     0     0     0     0     1
1682 1682     1     1     0     0     0     0     0     0     0
1683 1683     1     1     1     0     0     0     0     0     0
1684 1684     1     1     1     0     0     0     0     0     0
> cut.ch <- capHistSum(cutty,cols2use=-1)
> cut.ch$methodB.top
     i=1 i=2 i=3 i=4 i=5 i=6 i=7 i=8 i=9
j=1   NA  22   4   0   0   0   0   0   0
j=2   NA  NA  90   4   2   0   0   0   0
j=3   NA  NA  NA  37  13   1   0   0   0
j=4   NA  NA  NA  NA  43   2   1   0   0
j=5   NA  NA  NA  NA  NA  96   2   2   0
j=6   NA  NA  NA  NA  NA  NA  88  10   1
j=7   NA  NA  NA  NA  NA  NA  NA  40   4
j=8   NA  NA  NA  NA  NA  NA  NA  NA  13
j=9   NA  NA  NA  NA  NA  NA  NA  NA  NA
```

```
> cut.ch$methodB.bot
   i=1 i=2 i=3 i=4 i=5 i=6 i=7 i=8 i=9
m    0  22  94  41  58  99  91  52  18
u   89 330 198 192 201 271 199  82 122
n   89 352 292 233 259 370 290 134 140
R   89 352 292 233 259 370 290 134   0
```

11.3.1 Jolly-Seber Method

The Jolly-Seber method was introduced in Section 9.4.1 as a method to estimate abundance in an open population. When three or more samples of marked fish have been collected, the Jolly-Seber method can also be used to estimate the apparent survival (ϕ_i) from sampling event i to $i+1$. Derivations for the maximum likelihood estimators of ϕ_i (and associated standard errors) are throughly described in Pollock et al. (1990) and Krebs (1999) and will not be discussed in detail here. However, note that ϕ_{k-1} and ϕ_k, where k is the number of capture events, cannot be estimated.

Jolly-Seber calculations are performed by submitting the capHistSum() object to mrOpen() from **FSA**.[14] Jolly (1965) proposed a large-sample (i.e., using a normal distribution) method, but Manly (1984) provided an alternative method, to construct confidence intervals for ϕ_i. Jolly's method is used by default, but Manly's alternative may be used by including type="Manly" in mrOpen(). Pollock et al. (1990) suggested that most scientists are interested in ϕ_i as it relates to an individual fish, rather than as an average of all fish, and, thus, the variance of $\widehat{\phi}_i$ should include both sampling and individual variability. This was called the "full variance" for $\widehat{\phi}_i$ by Pollock et al. (1990) and is calculated by default with mrOpen().[15]

```
> cut.mr <- mrOpen(cut.ch)
```

The $\widehat{\phi}_i$ and associated confidence intervals are extracted from the mrOpen() object with summary() and confint() including parm="phi".

```
> cbind(summary(cut.mr,parm="phi"),confint(cut.mr,parm="phi"))
      phi phi.se phi.lci phi.uci
i=1 0.411  0.088   0.237   0.584
i=2 0.349  0.045   0.261   0.436
i=3 0.370  0.071   0.231   0.509
i=4 0.218  0.031   0.159   0.278
i=5 0.437  0.041   0.356   0.517
i=6 0.451  0.069   0.316   0.585
i=7 0.268  0.072   0.127   0.409
i=8   NA     NA      NA      NA
i=9   NA     NA      NA      NA
```

Thus, for 2003 (i.e., $i = 6$) the apparent survival to the following year was 0.451 with a 95% CI from 0.316 to 0.585.

11.3.2 Log-Linear Model or Cormack-Jolly-Seber Method

The log-linear model approach for estimating abundance as described by Cormack (1989) was introduced in Section 9.4.2. As described in that section, this method fits three types of models (*no "trap-effect," homogeneous "trap-effect,"* and *full "trap-effect"*) and then uses an information-theoretic approach (Burnham and Anderson 2002) to identify the "best" model. In addition to estimating abundance and the probability of capture (as shown in Section 9.4.2), ϕ_i can also be estimated from these models.

These models were fit and assessed for the Auke Lake Cutthroat Trout data in Section 9.4.2. These results are shown again below. Recall that the lower AIC value for the no trap-effect model suggests that there is not a substantial trap effect evident in these data.

```
> cut.op <- openp(cutty[,-1])
> cut.op$model.fit
               deviance     df       AIC
fitted model 86.33589     487   315.8965
> cut.op$trap.fit
                                     deviance     df       AIC
model with homogenous trap effect 85.19608     486   316.7567
model with trap effect            81.12745     481   322.6881
```

Estimates for the ϕ_i are extracted by appending $survivals to the openp() object. Approximate confidence intervals are obtained from normal theory. These extractions and calculations are illustrated below. Note that data.frame() is used because cut.op$survivals returns a matrix rather than a data.frame. Also note that qnorm() finds the value on a standard normal distribution that has the given proportional area lower.

```
> conf.level <- 0.95
> z <- qnorm(.5+conf.level/2)
> survs <- data.frame(cut.op$survivals) %>%
    mutate(phi.LCI=estimate-z*stderr,
           phi.UCI=estimate+z*stderr)
> survs
    estimate      stderr    phi.LCI    phi.UCI
1  0.4119850  0.08868670  0.2381623  0.5858078
2  0.3500535  0.04480100  0.2622451  0.4378618
3  0.3739616  0.07158820  0.2336513  0.5142719
4  0.2176769  0.03048458  0.1579282  0.2774255
5  0.4369129  0.04121960  0.3561240  0.5177019
6  0.4545608  0.06941919  0.3185017  0.5906199
7  0.2756015  0.07464255  0.1293048  0.4218982
8         NA          NA         NA         NA
```

Thus, for 2003 (i.e., row 6) the apparent survival to the following year was 0.455 (95% CI: 0.319, 0.591).

11.4 Mortality Components

In broad terms, mortality of fishes may be due to fishing (i.e., harvest) or natural causes (e.g., predation or senescence). It is difficult in most fisheries to separate A into these two components because some fish that were harvested during the year would have otherwise died naturally and vice versa. In contrast, in any given instant, mortality may be due to only one of these causes. Thus, Z can be partitioned into *instantaneous fishing mortality* (F) and *instantaneous natural mortality* (M) rates; that is, $Z = F + M$. More thorough explanations of these definitions and principles are in Miranda and Bettoli (2007) and Allen and Hightower (2010).

11.4.1 Estimates of M or F from Field Data

Miranda and Bettoli (2007, p. 256) described five methods to estimate M or F from various sources of field data — "(1) regression of Z as a function of fishing effort to estimate M, (2) catch-curve analysis to estimate M, (3) mark-recapture to estimate F, (4) direct census to estimate F, [and] (5) production modeling to estimate M." These methods require data that are generally difficult or costly to acquire and are subject to assumptions that are often difficult to meet. In addition, the resulting estimates of M or F are often highly uncertain because they are sensitive to measurement or estimation errors in the data. However, if adequate data exist, the methods are easy to implement as most either rely on simple linear regression models (see Section 7.3) or simple accounting and algebra. As no special purpose functions are required to fit these models in R, they will not be discussed further here.

Allen and Hightower (2010) and Hightower and Pollock (2013) summarized several models that use *passive* and *active* tagging data to estimate components of mortality. They define passive tagging as the tagging of fish with external tags and the subsequent receipt of reports regarding those tags from anglers who have caught the fish. In contrast, active tagging uses telemetry to remotely locate tagged fish at regular intervals. Models to estimate fishing and natural mortality from these data are complex and beyond the scope of this book. However, the models for estimating natural and fishing mortality from multi-year passive tagging data developed by Hoenig et al. (1998) and Jiang et al. (2007) have been implemented in `irm_h()` and `irm_cr()`, respectively, from **fishmethods**.

11.4.2 Meta-Analytic Estimates of M

The instantaneous rate of natural mortality (M) is highly correlated with various life history or abiotic factors. A number of researchers have exploited this relationship to develop simple models for predicting M from these other factors. This collection of methods has been called *indirect, empirical,* or *pragmatic alternatives for information-limited situations* (Kenchington 2014). Here, these methods are collectively referred to as *meta-analysis estimator equations.* Recent works by Kenchington (2014) and Then et al. (2015) thoroughly reviewed many of these methods and provided clear recommendations for use (or not). The reader is strongly urged to consult these resources before using the methods described below.

The two most commonly used relationships according to Then et al. (2015) are from Pauly (1980) and Hoenig (1983). Pauly (1980), through examination of results from 175 fish stocks, developed the relationship

$$M = 10^{-0.0066-0.279log_{10}(L_\infty)+0.6543log_{10}(K)+0.4634log_{10}(T)}$$

where K and L_∞ (cm) are parameters from the von Bertalanffy Growth Function (see Chapter 12) and T is the average temperature (in Celsius) experienced by the fish. Hoenig (1983) examined 130 stocks of fish, molluscs, and cetaceans, to obtain, for all stocks combined,

$$M = e^{1.44-0.982log_e(t_{max})}$$

or, for only the fish stocks,

$$M = e^{1.46-1.01log_e(t_{max})}$$

where t_{max} is the maximum age of the animal in the stock.

In an examination of these and several other meta-analysis estimation equations applied to over 200 fish stocks, Then et al. (2015) recommended that

$$M = 4.899t_{max}^{-0.916}$$

be used when possible, but that

$$M = 4.118K^{0.73}L_\infty^{-0.33}$$

could be used if t_{max} was not available.

The equations for these methods are simple and, thus, may be implemented by simply plugging values into the equations. For convenience, several meta-analysis estimating equations, including those mentioned above, have been coded in `metaM()` of **FSA**.[16] The first argument to `metaM()` is the name for the equation to be used. The five equations listed above are called `"PaulyL"`, `"Hoenig0"`, `"HoenigOF"`, `"tmax"`, and `"PaulyNoT"`, respectively. The other arguments to `metaM()` depend on which parameters are required by the method. The five equations above would use one or more of `T=`, `Linf=`, `K=`, and `tmax=`.

The examples below illustrate using `metaM()` to estimate M with just Pauly's equation and then all equations described above at once. The life history and abiotic information used in these examples is for Chesapeake Bay (Maryland and Virginia) Bay Anchovy as provided by Kenchington (2014).

```
> metaM("PaulyL",Linf=12.93,K=0.23,T=17)
[1] 0.685224
> tmp <- c("PaulyL","Hoenig0","HoenigOF","tmax","PaulyLNoT")
> metaM(tmp,Linf=12.93,K=0.23,T=17,tmax=3)
      method        M
1     PaulyL 0.6852240
2    Hoenig0 1.4349970
3   HoenigOF 1.4196375
4      tmax1 1.7030000
5  PaulyLNoT 0.6052296
```

11.5   Further Considerations

The catch curve methods (Chapman-Robson and regression) have been thoroughly reviewed in recent years. Specifically, Dunn et al. (2002) provided an excellent review of past examinations of the regression and Chapman-Robson methods. In addition, they examined precision and bias of these two methods in the face of stochastic errors in Z, number of fish at time of recruitment to the fishery, sampling, and aging. Overall, they found that the Chapman-Robson estimator was most precise and least biased; however, the advantage over the regression method declined somewhat with increasing amounts of stochastic error and increasing values of Z.

Smith et al. (2012) provided another study of precision between the regression and Chapman-Robson methods with the additional consideration of different definitions of the descending limb of the catch curve. They found that the Chapman-Robson method using all ages after the age with the peak catch and the weighted regression using all ages after and including the age with the peak catch performed similarly. However, they suggested using the Chapman-Robson method because it is based on a statistical foundation and has a generally smaller variance, whereas the weighting procedure in the regression method is *ad hoc*. Finally, Smith et al. (2012) concluded that the unweighted regression should not be used. These resources should be consulted for a more thorough explanation and discussion.

Notes

[1] Data manipulations in this chapter require functions from **magrittr** and **dplyr**, which are fully described in Chapter 2.

[2] Note that `exp()` raises its argument to the power of e.

[3] One common method used to "record time," especially annums, for fish is through ages estimated from calcified structures (e.g., otoliths).

[4] Smith et al. (2012) described Z and χ^2 statistical tests for identifying the peak of the catch curve. They noted, however, that there is little evidence that these tests have been implemented in the literature. Furthermore, in simulation studies, they found that these tests did not perform well and recommended that they not be used. Thus, these tests are not described further here.

[5] These data could have been entered into an external file and loaded into R as described in Section 2.1.3. In addition, these data often result from summarizing ages recorded for individuals or derived from an age-length key (see Chapter 5).

[6] The Chapman-Robson estimate of S is a maximum likelihood estimate derived from theory that assumes that the catches at each age on the descending limb of the catch-curve follow a geometric probability distribution. Proof of this is beyond the scope of this book.

[7] The Chapman-Robson method is also implemented in `agesurv()` of **fishmethods**.

[8] A vector of successive ages can be created with : as shown in the example. However, a vector of nonsuccessive ages can be created with `c()`. For example, if one wanted to use ages 2, 3, 4, and 6, then use `c(2,3,4,6)`.

[9] The $\widehat{Z}$ and $SE_{\widehat{Z}}$ originally proposed by Chapman and Robson (1960) is extracted by including `zmethod="original"` in the call to `chapmanRobson()`. The bias correction of Hoenig et al. (1983), not including the variance inflation factor of Smith et al. (2012), is extracted by including `zmethod="Hoenigetal"` in `chapmanRobson()`.

[10] The confidence interval for just one of the parameters can be returned by including that parameter within quotes in `parm=`.

[11] The unweighted catch curve method is also implemented in `agesurv()` of **fishmethods**.

[12] Note the use of `[[]]` in this code so that the results will not retain the original, and ultimately inappropriate, label. See Section 1.7.1.

[13] The mortality rates for each section can also be obtained by separately applying `catchCurve()` to data.frames created for fish from each section of the river. Separate applications of `catchCurve()` are also needed to produce variance estimates or confidence intervals for the separate values of Z.

[14] It is also possible to perform the Jolly-Seber calculations by manually entering (from previously summarized results) the top and bottom portions of the Method B table into separate matrices and then supplying those matrices as the first two arguments to `mrOpen()`. See the help documentation for `mrOpen` for an example of this functionality.

[15] The variance for $\widehat{\phi}_i$ that does not include individual variability is computed if `phi.full=FALSE` is used in `mrOpen()`.

[16] Similar methods have also been implemented in `M.empirical()` of **fishmethods**.

12

Individual Growth

The biomass of fish in a population that is closed to migration may change due to the addition of individuals through recruitment (Chapter 13), subtraction of individuals by mortality (Chapter 11), or growth of the individuals in the population. Individual fish growth is the result of the bioenergetic processes of anabolism and catabolism of tissue. These processes are related to relatively stable life history characteristics (e.g., long-lived species tend to grow more slowly than short-lived species) and relatively variable environmental factors such as food availability and temperature. Food availability, for example, is reflective of such things as habitat suitability, competition (intra- and inter-specific), and water temperature. Thus, growth is a process that synthesizes a wide variety of abiotic and biotic factors.

Individual growth is treated as an increase in either length or weight (here-after, just length) with increasing age. Several functions have been used to model the mean length-at-age of fishes. These functions are core components of surplus production, yield-per-recruit, or statistical catch-at-age models, which form the basis for many management recommendations.

Most models of fish growth are nonlinear. Thus, this chapter also serves as an introduction to fitting nonlinear models. Note, however, that the focus here is on how to fit, assess, and interpret nonlinear regression models in R, whereas the statistical theory underlying them is only briefly discussed.[1]

Required Packages for This Chapter

Functions used in this chapter require loading the packages shown below.[2]

```
> library(FSA)
> library(magrittr)
> library(dplyr)
> library(nlstools)
```

12.1   Data Requirements

Historically, growth functions were fit to mean length-at-age data. However, to fit models with appropriate estimates of variability, length (L) measurements and age (t) estimates from individual fish are required. For example, researchers recorded the total length (mm), age (as estimated from thin-sectioned otoliths), and sex of 140 Black Drum captured in 2001 from Virginia waters of the Atlantic Ocean. These data are available in *BlackDrum2001.csv*. For the purposes of this chapter, a single 51-year old fish, immature or unknown sex fish, and several unused variables were removed from the data. The first part of this chapter will use the data from only male Black Drum (i.e., bdm), whereas the latter part will compare growth models between male and female Black Drum (i.e., bdmf).

```
> bdmf <- read.csv("BlackDrum2001.csv") %>%
    filterD(otoage<50,sex %in% c("male","female")) %>%
    select(-c(spname,day,weight))
> headtail(bdmf,n=2)
    year agid month     tl    sex otoage
1   2001    1     4  787.5   male      6
2   2001    2     5  700.0   male      5
107 2001  127     6  530.0   male      3
108 2001  134    10  255.0 female      0
> bdm <- filterD(bdmf,sex=="male")
```

Growth functions may be fit to length-at-age data back-calculated from measurements on calcified structures. However, the methods described in this chapter are not appropriate for those data because the multiple observed lengths for individual fish are not independent. These data may be analyzed with repeated measure (Jones 2000) or mixed model (Vigliola and Meekan 2009; Escati-Penaloza et al. 2010) approaches.[3] These data or types of analyses are not considered further here.

12.2   Growth Functions

Mean length-at-age has been modeled by a variety of functions including the exponential, logistic, and polynomial (Ricker 1975), as well as the specific functions of von Bertalanffy (1938), Gompertz (1825), Richards (1959), Schnute (1981), and Schnute and Richards (1990). Each of the specific functions is nonlinear and, thus, requires nonlinear model fitting methods (Section 12.3) to fit the function to data.

   The von Bertalanffy growth function (VBGF) is by far the most commonly used growth function in fisheries analyses (Haddon 2011), despite several criticisms (e.g., Knight 1968; Roff 1980; Katsanevakis and Maravelias 2008; Haddon 2011). The most common representation of the VBGF, hereafter called the "Typical VBGF," was described by Beverton and Holt (1957) and is

$$E[L|t] = L_\infty \left(1 - e^{-K(t-t_0)}\right) \tag{12.1}$$

where $E[L|t]$ is the mean length-at-age t and L_∞, K, and t_0 are parameters to be estimated.

The three parameters in the Typical VBGF have strict meanings that require careful interpretation (Haddon 2011). First, L_∞ is the maximum *mean* length; it is **not** the maximum length of an individual (Figure 12.1). In other words, it is possible that an individual fish is longer than L_∞. Second, K relates how quickly the function approaches L_∞. Specifically, the time required to grow halfway between any length and L_∞ is $\frac{log_e(2)}{K}$. It must be noted that K is **not** a *growth rate* (Ricker 1975), as the units are inverse time (e.g., $\frac{1}{year}$) rather than a length increment per unit time (e.g., $\frac{mm}{year}$) as required by a growth rate. Third, t_0 is the "x-intercept" and is needed for model fitting. The t_0 parameter does not have a biological interpretation because $L = 0$ does not exist.

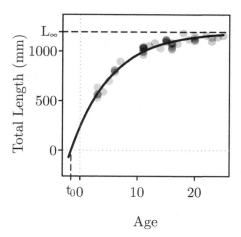

FIGURE 12.1. The von Bertalanffy growth function (solid black curved line) fit to male Black Drum (gray points). Representations of L_∞ and t_0 are shown. The plot was restricted to age 25 and younger fish to better illustrate the parameters.

Methods in this chapter are illustrated with the VBGF, but the methods are easily applied to any other growth function.[4]

12.3   Fitting Nonlinear Regressions

Many models in fisheries and ecological research are nonlinear. Some of these models can be transformed to a linear scale where traditional linear regression methods (Section 7.3) can be used. Other models, such as the VBGF and stock-recruit functions in Section 13.1, cannot be transformed to a linear model and, thus, are fit using nonlinear regression techniques.

The usual nonlinear regression model for individuals is expressed as

$$y_i = m(\vec{X}; \vec{\theta}) + \epsilon_i \qquad (12.2)$$

where y is a response variable, $\vec{X}$ is one or more explanatory variables, $\vec{\theta}$ is one or more parameters, $m(\bullet)$ is a nonlinear mean function (e.g., Equation (12.1)) that relates the response variable to the explanatory variable(s), and ϵ_i are individual additive errors around $m(\bullet)$ (Fox and Weisberg 2010). The residual sum-of-squares (RSS) and likelihood functions for these models are nonlinear and, thus, closed-form solutions for the parameters that minimize the RSS or maximize the likelihood do not exist. Thus, iterative numerical approximation solutions implemented with computer optimization algorithms are required to estimate the parameters.

Motulsky and Ransnas (1987) provided a nice description of the concept of fitting nonlinear models to data. Their description relies on visualizing the RSS surface in a simple example that has two parameters (Figure 12.2). In general, the algorithm begins with the user supplying a set of starting values for the model parameters (discussed further below). The algorithm then selects, based on criteria that depend on the choice of algorithm, another set of parameter values that are generally "downhill" from the starting values. This process continues with each iteration providing a set of parameter values that are closer to the values that minimize the RSS. Most algorithms continue until the improvement in lowering the RSS from one iteration to the next is considered negligible. The algorithm proceeds similarly for models with more parameters, though the surface would be more than three-dimensional and, thus, not able to be visualized.

12.3.1   Starting Values

There are at least three methods to identify reasonable starting values (denoted below with a ~ over the parameter) for Equation (12.1). Typically only one of these methods would be used, but, on occasion, one method might produce inadequate starting values (see Section 12.5), whereas another method would not. Thus, as no one method can be recommended for use in all situations, each of these methods is illustrated below.

First, Ford (1933) and Walford (1946) showed that a plot of L_{t+1} versus L_t is linear with a slope equal to e^{-K}. Additionally, the point on this line

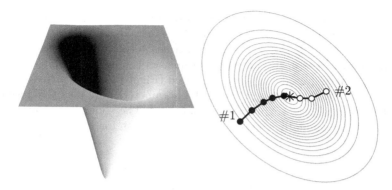

FIGURE 12.2. A schematic three-dimensional (Left) and equal contours (Right) plot that represents the RSS surface for two parameters. The nonlinear regression algorithm begins at user-defined starting values and iteratively "moves downhill" to minimize the RSS. Two example traces of this process are shown on the contour plot.

where $L_{t+1} = L_t$ is L_∞. From this, $\tilde{L}_\infty = \frac{intercept}{1-slope}$ and $\tilde{K} = -log(slope)$, where *intercept* and *slope* are the y-intercept and slope from fitting the linear regression of L_{t+1} on L_t. With $\tilde{L}_\infty$ and observed values for one age and mean length pair (i.e., $(\acute{t}, \bar{L}_i)$), Equation (12.1) is solved for t_0 to obtain $\tilde{t}_0 = \acute{t} * log\left(\frac{\tilde{L}_\infty - \bar{L}_i}{\tilde{L}_\infty}\right)$. Alternatively, the root closest to zero from a second-degree polynomial fit to the mean length-at-age data (i.e., find where the polynomial function crosses the age axis closest to 0) may be used for $\tilde{t}_0$.

These methods for generating starting values are implemented in vbStarts() of **FSA**. The first argument to vbStarts() is a formula of the form len~age, where len and age generically represent the variables containing the lengths and ages for individual fish in the data.frame in data=. By default, $\tilde{L}_\infty$ and the mean length for the youngest observed age is used to produce $\tilde{t}_0$. A $\tilde{t}_0$ based on the polynomial regression method is obtained by including meth0="poly". Include plot=TRUE to see how the VBGF using the obtained starting values fits the observed data. The list of starting values returned by vbStarts() must be assigned to an object for later use.

```
> ( svTyp <- vbStarts(tl~otoage,data=bdm) )
$Linf
[1] 1192.692

$K
[1] 0.1924053

$t0
[1] -0.4010652
```

Starting values for the VBGF are also obtained by extracting parameters from a visual fit of the VBGF to the observed data (Ritz and Streibig 2008; Bolker et al. 2013). For example, a scatterplot with Equation (12.1) superimposed on the data, but with parameters connected to slider bars that adjust the superimposed model when moved (Figure 12.3) is constructed with `dynamicPlot=TRUE` in `vbStarts()`.[5] The parameters may be altered with the slider bars until an approximate fit is obtained. The slider bar values at that point are then manually entered into a named list for later use as starting values.

```
> vbStarts(tl~otoage,data=bdm,dynamicPlot=TRUE)
> svTyp <- list(Linf=1193,K=0.13,t0=-2.0)
```

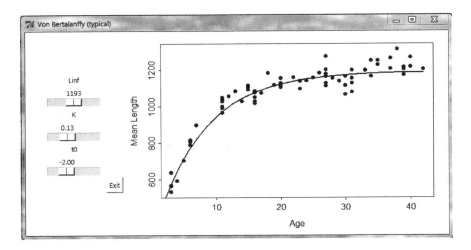

FIGURE 12.3. Example of using dynamic plots (e.g., `dynamicPlot=TRUE` in `vbStarts()`) to find an approximate fit of the Typical von Bertalanffy growth function to male Black Drum.

Finally, experience suggests that values of t_0 and K tend to be near 0 and 0.3, respectively. Additionally, L_∞ is often near the maximum observed length in the sample. Thus, $\tilde{L}_\infty = max(L_t)$, $\tilde{K} = 0.3$, and $\tilde{t}_0 = 0$ work reasonably well for many species. The maximum length is found by submitting the vector of observed lengths to `max()` and using `na.rm=TRUE` to remove any missing lengths that may exist in the vector. These values are entered into a named list for later use.

```
> svTyp <- list(Linf=max(bdm$tl,na.rm=TRUE),K=0.3,t0=0)
```

12.3.2   Model Fitting

Subsequent analysis of model results is more efficient if the VBGF is entered into R as an object or as a function. When working with a single group of individuals, a function is more flexible and is illustrated in this section. When comparing VBGFs fit to separate groups of individuals, an object is easier to use, as will be illustrated in Section 12.4.

User-defined functions are created with function() as described in Section 1.8.2. A user-defined function for the VBGF should take age as the first argument, three parameters as the next three arguments, and return a length predicted from the VBGF. For example, the code below creates a function called vbTyp() that has age, Linf, K, and t0 as arguments and returns a predicted length according to Equation (12.1). Note that exp() is used to raise its argument to the power of *e*.

```
> vbTyp <- function(age,Linf,K,t0) Linf*(1-exp(-K*(age-t0)))
```

As a test, the function is used to predict the mean length for an age-3 fish.

```
> vbTyp(3,Linf=1200,K=0.13,t0=-2.0)
[1] 573.5451
```

A function for the VBGF can also be created with vbFuns() from **FSA**. No arguments are required in vbFuns() when using Equation (12.1). The result of vbFuns() is assigned to an object which then becomes the name of the created function. The function created with vbFuns() is more complicated than what was illustrated above because it allows values for all three parameters to be given in the first argument, which will be useful in Section 12.3.4.

```
> vbTyp <- vbFuns()
> vbTyp(3,Linf=1200,K=0.13,t0=-2.0)     # this still works
[1] 573.5451
> vbTyp(3,Linf=c(1200,0.13,-2.0))       # but, now, so does this
[1] 573.5451
```

Parameters for the VBGF are estimated with nls(), which requires a model formula as the first argument.[6] In the case with vbTyp(), this formula will have the form len~vbTyp(age,Linf,K,t0), where len and age are generic names for the length and age variables in the data.frame given to data=. Note that Linf, K, and t0 must be listed as shown and in the same order as in vbTyp(). Finally, starting values for the parameters must be given as a named list (i.e., as created with vbFuns() in Section 12.3.1) to start=.

```
> fitTyp <- nls(tl~vbTyp(otoage,Linf,K,t0),data=bdm,start=svTyp)
```

12.3.3 Parameter Estimates

Parameter estimates are extracted from the `nls()` object with `coef()`.

```
> coef(fitTyp)
         Linf            K           t0
 1196.7184461    0.1418279   -1.5942873
```

Thus, $\widehat{L}_\infty{=}1197$, $\widehat{K}{=}0.142$, and $\widehat{t}_0{=}{-}1.59$.

Confidence intervals for parameters in nonlinear models should be computed from profile likelihood or bootstrap methods (as described in detail in Ritz and Streibig (2008)). Profile likelihood confidence intervals are extracted from the `nls()` object with `confint()`.[7]

```
> confint(fitTyp)
Waiting for profiling to be done...
            2.5%         97.5%
Linf 1177.9880088 1217.3520908
K       0.1214678    0.1649841
t0     -2.6594327   -0.7343047
```

Bootstrapping (nonparametric) a nonlinear model is best done with `nlsBoot()` from **nlstools** (Baty et al. 2015). This function takes the `nls()` object as its only required argument. By default, 999 data sets of the same number of age and lengths are generated with replacement from the original age and length data, the same model used in the `nls()` object is fit to each data set, and parameters are extracted from each model fit and returned in the `coefboot` object of the assigned object.[8]

```
> bootTyp <- nlsBoot(fitTyp)
> headtail(bootTyp$coefboot,n=2)
             Linf         K          t0
[1,]     1219.916 0.1236133 -2.173670
[2,]     1199.656 0.1328963 -2.020926
[998,]   1203.133 0.1273796 -2.039798
[999,]   1195.790 0.1487650 -1.166369
```

Bootstrapped 95% confidence intervals (i.e., the 2.5 and 97.5 percentiles of the bootstrapped estimates) are extracted from the `nlsBoot()` object for each parameter with `confint()`. Including `plot=TRUE` in `confint()` will produce histograms of the parameter estimates from the bootstrapped samples augmented with a depiction of the confidence intervals (Figure 12.4).

```
> confint(bootTyp,plot=TRUE)
             95% LCI      95% UCI
Linf 1180.0722726 1216.6023027
K       0.1236074    0.1622241
t0     -2.5745127   -0.8522599
```

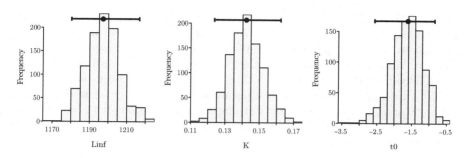

FIGURE 12.4. Histograms of parameter estimates for the von Bertalanffy growth function, with depictions of the 95% confidence intervals, from bootstrapped samples for male Black Drum.

The profile likelihood and bootstrap confidence intervals may differ dramatically if the profile (not shown) is greatly curved or if the histogram of bootstrapped estimates is highly skewed (Figure 12.4). In general, the bootstrapped confidence intervals likely have better coverage.

More results for the model fit are extracted from the nls() object with summary(), which may include correlation=TRUE for nls() objects.

```
> summary(fitTyp,correlation=TRUE)

Formula: tl ~ vbTyp(otoage, Linf, K, t0)

Parameters:
        Estimate Std. Error t value Pr(>|t|)
Linf 1196.71845    9.41723 127.078  < 2e-16
K       0.14183    0.01005  14.114  < 2e-16
t0     -1.59429    0.44941  -3.548 0.000694

Residual standard error: 45.6 on 71 degrees of freedom

Correlation of Parameter Estimates:
   Linf  K
K  -0.72
t0 -0.51  0.89

Number of iterations to convergence: 6
Achieved convergence tolerance: 4.843e-06
```

A summary of parameter estimates appears under "Parameters:", with the results from coef() repeated under "Estimate". Standard errors for each parameter and t-tests and corresponding p-values for whether each parameter is

statistically equal to zero or not are also shown here. These hypothesis tests
are generally not of interest in VBGF fits and depend upon linear approximations to the model (Ritz and Streibig 2008), which may be quite poor with
VBGFs. More appropriate hypothesis tests and procedures are discussed below. The variability about the model is shown after "Residual standard error:"
(=45.6). Correlations between each parameter in the model are in the matrix
under "Correlation of Parameter Estimates:". Correlations between parameters are high for most VBGF fits (see Section 12.5). Algorithm performance
information is shown in the last two lines.

12.3.4 Predictions

Predicted lengths-at-age are obtained from the results of the fitted VBGF in
a variety of ways. First, `predict()` is used with the `nls()` object as the first
argument and a data.frame of ages at which to compute the predicted lengths
as the second argument (Section 7.3.1). The data.frame must have a variable
named exactly as the age variable was named in the `nls()` call (i.e., `otoage`
in this example).

```
> nd <- data.frame(otoage=c(3,10,20,30,40,42))
> predict(fitTyp,nd)
[1]   572.9742   965.5966 1140.7568 1183.1684 1193.4376 1194.2479
```

Second, the user-created model function (e.g., `vbTyp()`) is used with a vector of ages at which to make the predictions as the first argument and the
parameter estimates from `coef()` as the second argument (this assumes that
the user-created function allows values for all three arguments in the first
argument as shown in Section 12.3.2).

```
> vbTyp(c(3,10,20,30,40,42),coef(fitTyp))
[1]   572.9742   965.5966 1140.7568 1183.1684 1193.4376 1194.2479
```

The two methods produce the same results, but neither method produces
estimates of variability associated with the prediction. Confidence intervals for
the predicted mean length can, however, be constructed from the bootstrap
results by predicting a mean length at a given age for each bootstrap sample
and then finding the 2.5 and 97.5 percentiles (for a 95% confidence interval)
of those predictions.[9]

The parameter estimates from each bootstrap sample are, as illustrated
above, in the `coefboot` object of the `nlsBoot()` object. The predicted mean
length at age-3, for example, for one of the bootstrap samples is computed by
supplying one row from `bootTyp$coefboot` to the VBGF function.

```
> vbTyp(3,bootTyp$coefboot[1,])
[1]  576.3637
```

The `apply()` function allows this calculation to be efficiently repeated for **all** rows as described in Section 1.9.2. The first argument to `apply()` is the matrix that contains the parameter estimates from each bootstrap sample. The function in `FUN=` **must** allow the values for the three parameters to be sent in one argument (i.e., like `vbTyp()` as created with `vbFuns()` in Section 12.3.2). The last argument is the age value used for the prediction set equal to the first argument of `vbTyp()` (note that the functions created with `vbFuns()` use `t=` as the "age" argument).

```
> p3Typ <- apply(bootTyp$coefboot,MARGIN=1,FUN=vbTyp,t=3)
> p3Typ[1:6]    # show predictions for first 6 bootstrap samples
[1] 576.3637 584.0989 536.4806 569.6764 587.4837 580.0032
```

The 95% confidence interval is then the 2.5 and 97.5 percentiles (also called quantiles) from the vector of predicted values as found with `quantile()`.

```
> quantile(p3Typ,c(0.025,0.975))
    2.5%    97.5%
533.7009 610.8455
```

Thus, one is 95% confident that the mean length for age-3 Black Drum is between 534 and 611 mm.

12.3.5   Visualizing the Model Fit

A plot with the best-fit VBGF superimposed (Figure 12.5) is constructed with `lines()` as described in Sections 3.5.1 and 7.3.2. One difference in the code below is that the ranges for the x- and y-axes are set with `xlim=` and `ylim=` in `plot()` because the VBGF was modeled outside the range of the age data.

```
> x <- seq(0,42,length.out=199)        # ages for prediction
> pTyp <- vbTyp(x,Linf=coef(fitTyp))   # predicted lengths
> xlmts <- range(c(x,bdm$age))
> ylmts <- range(c(pTyp,bdm$tl))
> plot(tl~otoage,data=bdm,xlab="Age",ylab="Total Length (mm)",
       xlim=xlmts,ylim=ylmts,pch=19,col=rgb(0,0,0,1/3))
> lines(pTyp~x,lwd=2)
```

Adding confidence bands[10] to the plot follows the same general principles shown in Section 7.3.2. However, creating vectors of lower and upper confidence band values is more work here because the intervals must be created one age at a time from the bootstrap results. Thus, code that is similar to that described in Section 12.3.4 for age-3 fish is placed within a loop that will cycle through all ages in `x` (from above). The confidence interval values computed in this loop are then plotted against `x` to make confidence bands.

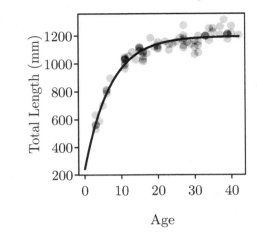

FIGURE 12.5. Length versus age with superimposed best-fit von Bertalanffy growth function for male Black Drum.

The process of constructing a loop was discussed in Section 1.9.1. Two numeric vectors (LCI and UCI) that are the same length as the number of ages in x are needed to hold the lower and upper confidence interval values, respectively.

```
> LCI <- UCI <- numeric(length(x))
```

The for() function is then used to loop through each age in x. At each step in the loop, the predicted length and lower and upper confidence interval values are computed (as shown above for one age) for an age in x and stored in the *i*th position of the LCI and UCI vectors.

```
> for(i in 1:length(x)) {
    tmp <- apply(bootTyp$coefboot,MARGIN=1,FUN=vbTyp,t=x[i])
    LCI[i] <- quantile(tmp,0.025)
    UCI[i] <- quantile(tmp,0.975)
  }
```

The values in LCI and UCI are then added to the plot with lines() to produce Figure 12.6. Note that the ylmts object from above was modified to adjust for the values in LCI and UCI. The confidence bands for the male Black Drum are very narrow due to the lack of individual variability in these data and the very tight fit of the VBGF.

```
> ylmts <- range(c(pTyp,LCI,UCI,bdm$tl))
> plot(tl~otoage,data=bdm,xlab="Age",ylab="Total Length (mm)",
      xlim=xlmts,ylim=ylmts,pch=19,col=rgb(0,0,0,1/3))
> lines(pTyp~x,lwd=2)
> lines(UCI~x,lwd=2,lty="dashed")
> lines(LCI~x,lwd=2,lty="dashed")
```

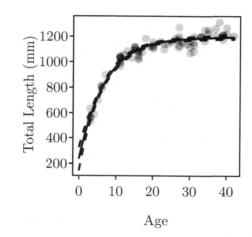

FIGURE 12.6. Length versus age with superimposed best-fit von Bertalanffy growth function and 95% confidence bands (dashed lines) for male Black Drum.

12.3.6 Assumptions

The nonlinear regression method assumes that the specified model fits the data (i.e., the mean function (e.g., Equation (12.1)) is correct) and that the measurement errors are independent, are normally distributed, and have equal variance (i.e., homoscedastic). The assumptions about the measurement errors are not evident until Equation (12.1) is expressed for the length of an individual fish, L_i, rather than a mean length, as

$$L_i = L_\infty \left(1 - e^{-K(t_i - t_0)}\right) + \epsilon_i \tag{12.3}$$

where the ϵ_i represents the deviance of the ith individual from the model. Under the assumptions, $\epsilon \sim N(0, \sigma)$.

As with linear models (Section 7.3.3), the independence assumption depends on how the data were collected and is not addressed further here. Statistical tests exist for checking the other assumptions but these assumptions are often better assessed by interpreting a residual plot and a histogram of residuals. These plots are constructed with `residPlot()` exactly as shown for linear models (see Section 7.3.3).[11]

```
> residPlot(fitTyp)
```

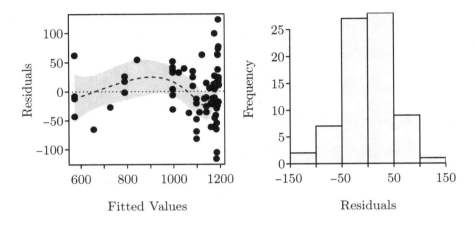

FIGURE 12.7. Residual plot (left) and histogram of residuals (right) from fitting the Typical von Bertalanffy growth function to the male Black Drum.

The homoscedasticity assumption assumes that variability in fish length will remain constant across all observed ages of the fish. Length variability will very likely increase with age for most fish species and, thus, the homoscedasticity assumption is often violated. The slight funnel shape in the residual plot for male Black Drum (Figure 12.7) suggests a slight heteroscedasticity.

When heteroscedasticity is present, the errors may be modeled as multiplicative rather than additive. Thus, with an assumption of multiplicative errors, Equation (12.3) becomes

$$L_i = L_\infty \left(1 - e^{-K(t_i - t_0)}\right) e^{\epsilon_i} \tag{12.4}$$

Natural logarithms of both sides of Equation (12.4) gives

$$log_e(L_i) = log_e \left(L_\infty \left(1 - e^{-K(t_i - t_0)}\right)\right) + \epsilon_i \tag{12.5}$$

which reveals that the error structure is additive on the log scale. Thus, model parameters are estimated using the techniques described previously with the modification that the length variable and the right-hand side of the formula are log-transformed (with `log()`).

```
> bdm %<>% mutate(logTL=log(tl))      # add log TL variable to bdm
> fitTypM <- nls(logTL~log(vbTyp(otoage,Linf,K,t0)),
                    data=bdm,start=svTyp)
> residPlot(fitTypM)
```

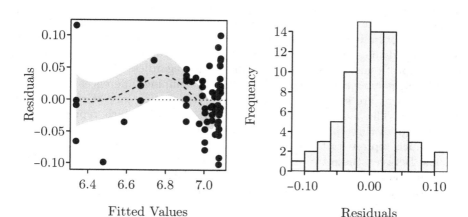

FIGURE 12.8. Residual plot (left) and histogram of residuals (right) from fitting the Typical von Bertalanffy growth function to male Black Drum, assuming multiplicative errors.

In this case, fitting the multiplicative errors model improved the homoscedasticity of the errors somewhat (Figure 12.8). Both models suggest a slight "waviness" to the residuals that may suggest slight differences in growth among year-classes or periods of years. Given that the improvement in how the model meets the assumptions is not substantial, the additive (i.e., untransformed) model is likely adequate.

12.4   Among Group Statistical Comparisons

Fisheries scientists often compare growth between two or more groups. To make this comparison, a factor variable that identifies to which group an individual belongs is required. Internally, R will convert this factor variable to one or more dummy variables as described in Section 7.4. A method for comparing two groups is illustrated below for male and female Black Drum using the bdmf data.frame constructed at the beginning of the chapter.

12.4.1 Family of Models

Comparing VBGFs between groups requires the fitting of as many as eight models.[12] All but one of these models represents a difference between groups for at least one of the parameters. Thus, parameter notation must be adjusted to indicate to which group the parameter pertains. A number in square brackets is appended to the parameter to denote groups. For example, $L_\infty[1]$ is the asymptotic mean length for individuals in group 1 and $t_0[2]$ is the "x-intercept" for individuals in group 2. A parameter without a bracketed number is considered the same (i.e., not different) among groups. For example, L_∞ is the asymptotic mean length for all individuals, regardless of group.[13]

The eight models to be considered when examining differences in VBGFs among groups are shown, along with abbreviations for ease of reference, in Table 12.1. The most complex model is $\{L_\infty, K, t_0\}$, which represents the case where all three parameters differ among groups. The simplest model is $\{\Omega\}$, which represents the case where none of the parameters differ among groups. Between these two extremes are three models where two parameters differ among groups — $\{L_\infty, K\}$, $\{L_\infty, t_0\}$, and $\{K, t_0\}$ — and three models where only one parameter differs among groups — $\{L_\infty\}$, $\{K\}$, and $\{t_0\}$.

TABLE 12.1. The family of models considered when examining differences in VBGF among groups. Parameters are defined in the text, but note that *group* is 1 if in "group 1," 2 if in "group 2," and so on for the number of groups compared. The abbreviations denote which parameters differ among groups for that model. No parameters differ among groups for the Ω model.

Abbreviation	Model
$\{L_\infty, K, t_0\}$	$E[L\|t] = L_\infty[group] \left(1 - e^{-K[group](t - t_0[group])}\right)$
$\{L_\infty, K\}$	$E[L\|t] = L_\infty[group] \left(1 - e^{-K[group](t - t_0)}\right)$
$\{L_\infty, t_0\}$	$E[L\|t] = L_\infty[group] \left(1 - e^{-K(t - t_0[group])}\right)$
$\{K, t_0\}$	$E[L\|t] = L_\infty \left(1 - e^{-K[group](t - t_0[group])}\right)$
$\{L_\infty\}$	$E[L\|t] = L_\infty[group] \left(1 - e^{-K(t - t_0)}\right)$
$\{K\}$	$E[L\|t] = L_\infty \left(1 - e^{-K[group](t - t_0)}\right)$
$\{t_0\}$	$E[L\|t] = L_\infty \left(1 - e^{-K(t - t_0[group])}\right)$
$\{\Omega\}$	$E[L\|t] = L_\infty \left(1 - e^{-K(t - t_0)}\right)$

Each model in Table 12.1 is a nested subset of any model that shares **all** of the parameters in its abbreviation. For example, $\{L_\infty, K\}$ is a nested subset of only $\{L_\infty, K, t_0\}$, but $\{L_\infty\}$ is a nested subset of $\{L_\infty, t_0\}$, $\{L_\infty, K\}$, and $\{L_\infty, K, t_0\}$. By definition, $\{\Omega\}$ is a nested subset of all other models in Table

12.1. The nested relationships among these models allows use of likelihood ratio and extra sum-of-squares tests as will be illustrated in Section 12.4.2.

In Section 12.3.2, the VBGF was expressed as a function. The VBGF can also be expressed as an object, which is simpler to declare and to read, but does not allow efficient bootstrapping of predictions (see Section 12.3.4). The simplicity outweighs the bootstrapping concerns when comparing models and, thus, the eight models in Table 12.1 are entered into objects below for analysis in Section 12.4.2. Note in these declarations that the variable in square brackets is treated like a subscript and that `tl`, `otoage`, and `sex` are the names of the variables in the `bdmf` data.frame and may be different in other data.frames.

```
> vbLKt <- tl~Linf[sex]*(1-exp(-K[sex]*(otoage-t0[sex])))
> vbLK  <- tl~Linf[sex]*(1-exp(-K[sex]*(otoage-t0)))
> vbLt  <- tl~Linf[sex]*(1-exp(-K*(otoage-t0[sex])))
> vbKt  <- tl~Linf*(1-exp(-K[sex]*(otoage-t0[sex])))
> vbL   <- tl~Linf[sex]*(1-exp(-K*(otoage-t0)))
> vbK   <- tl~Linf*(1-exp(-K[sex]*(otoage-t0)))
> vbt   <- tl~Linf*(1-exp(-K*(otoage-t0[sex])))
> vb0   <- tl~Linf*(1-exp(-K*(otoage-t0)))
```

12.4.2 Model Fitting

Fit and Assess General Model

Analysis of the multiple nonlinear models begins by checking the assumptions for the most complex model $\{L_\infty, K, t_0\}$. If the assumptions are met with this model, then they should also be met for all nested models that do not significantly differ.

Fitting $\{L_\infty, K, t_0\}$ requires starting values for L_∞, K, and t_0 for **each** group. The starting values list consists of three named vectors of starting values for each parameter for each group. If the groups do not differ dramatically, then identical starting values can be defined for both groups.[14] If this is the case, then it is most efficient to first enter or derive a single list of the three common starting values as shown in Section 12.3.1.

```
> ( sv0 <- vbStarts(tl~otoage,data=bdmf) )
$Linf
[1] 1193.483

$K
[1] 0.2143151

$t0
[1] -1.121557
```

These single starting values are then expanded to starting values for each group with `Map()`, using `rep()` (for replicate) as the first argument. The second argument to `Map()` is the list of values to be replicated and the third argument is a numeric vector that describes how many times each value in the list should be repeated. For example, the following code replicates the starting values in `sv0` twice each (i.e., for the two sexes) to form starting values for $\{L_\infty, K, t_0\}$.

```
> ( svLKt <- Map(rep,sv0,c(2,2,2)) )
$Linf
[1] 1193.483 1193.483

$K
[1] 0.2143151 0.2143151

$t0
[1] -1.121557 -1.121557
```

These starting values and the `vbLKt` object declared in Section 12.4.1 are then used to fit $\{L_\infty, K, t_0\}$. A residual plot and histogram of residuals (Figure 12.9) suggest that $\{L_\infty, K, t_0\}$ fits the data appropriately, with only slight heteroscedasticity and lack-of-fit, but approximately normal residuals. This model could be fit with multiplicative errors as described previously, but that did not substantively improve the fit (results not shown). Thus, the additive errors (i.e., untransformed) model is used here.

```
> fitLKt <- nls(vbLKt,data=bdmf,start=svLKt)
> residPlot(fitLKt,col=rgb(0,0,0,1/3))
```

Identify Any Differences

If there is no significant difference between $\{L_\infty, K, t_0\}$ and $\{\Omega\}$, then $\{\Omega\}$ fits adequately and there is evidence for no difference in any of the parameters among the groups. In this case, no other models need to be considered. However, if there is a significant difference between $\{L_\infty, K, t_0\}$ and $\{\Omega\}$, then $\{L_\infty, K, t_0\}$ fits better and there is evidence that at least one of the parameters differs between at least one pair of groups. In this case, other models need to be considered to isolate which parameters and groups differ. This section describes how to compare $\{L_\infty, K, t_0\}$ and $\{\Omega\}$. The next section will describe how to incorporate the other models into a followup analysis.

Differences in nonlinear models may be identified with likelihood ratio tests (Kimura 1980), extra sums-of-squares tests (Ritz and Streibig 2008), or an information criterion (IC) measure (Burnham and Anderson 2002). The analytical details of these tests will not be described here. In general, the better of two models is the one with a significantly greater likelihood, smaller

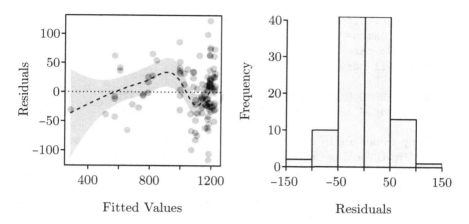

FIGURE 12.9. Residual plot (left) and histogram of residuals (right) from fitting the most general von Bertalanffy growth function to male and female Black Drum.

residual sum-of-squares, or lower IC value. The likelihood ratio tests and extra sums-of-squares tests are implemented in a sequential manner that is described next. The IC measures are generally computed on the entire suite of models considered and, thus, will not be described until later in this section.

The two models to compare must be fit and the results assigned to separate objects. For this comparison, $\{L_\infty, K, t_0\}$ was fit in the previous section and assigned to fitLKT and $\{\Omega\}$ is fit and assigned to fit0 below (note the use of the starting values in sv0 that were created previously).

```
> fit0 <- nls(vb0,data=bdmf,start=sv0)
```

The likelihood ratio and extra sum-of-squares tests for comparing two models are computed with lrt() and extraSS() from **FSA**, respectively. Both functions require the object with the simplest (i.e., the nested) model as the first argument and the object with the more complex model in com=. Descriptive names for the simple model(s) and the complex model may be given in sim.names= and com.name=, respectively.

```
> lrt(fit0,com=fitLKt,com.name="All pars differ",
      sim.names="No pars differ")
Model 1: No pars differ
Model A: All pars differ

    Df0  logLik0 DfA  logLikA Df  logLik Chisq Pr(>Chisq)
1vA 105 -565.8258 102 -561.2295  3 -4.5964 9.1927    0.02684
```

```
> extraSS(fit0,com=fitLKt,com.name="All pars diff",
      sim.names="No pars diff")
Model 1: No pars diff
Model A: All pars diff

   Df0   RSS0 DfA   RSSA Df       SS      F  Pr(>F)
1vA 105 224698 102 206363   3    18334 3.0207 0.03319
```

The p-values are in the last column of results from these functions (i.e., under "Pr(>Chisq)" or "Pr(>F)," respectively). The likelihood ratio ($\chi^2 = 9.2$, $p = 0.0268$) and extra sum-of-squares test ($F = 3.0$, $p = 0.0332$) both suggest a significant difference between $\{L_\infty, K, t_0\}$ and $\{\Omega\}$. Thus, there is evidence that there is some difference in the VBGF parameters between female and male Black Drum. The methods of the next section are used to identify which parameters differ between the two sexes.

Finding the Best Subset Model

Once it has been determined that there is some difference in parameters between at least one pair of groups (i.e., $\{L_\infty, K, t_0\}$ is preferred over $\{\Omega\}$), then the other six models are examined in a hierarchical manner to determine which is the most parsimonious model that still adequately fits the data. In the first step, the most complex nested subsets of $\{L_\infty, K, t_0\}$ are compared to $\{L_\infty, K, t_0\}$. Any nested model that is not statistically different from $\{L_\infty, K, t_0\}$ is considered "better" than $\{L_\infty, K, t_0\}$ because it fits equally (statistically) well, but is more parsimonious. If two nested models are better than $\{L_\infty, K, t_0\}$, then the one with the greatest log-likelihood or smallest RSS (depends on which metric you have chosen to use) is chosen as the "best" of the nested models. In the second step, the most complex models that are nested subsets of the best model from the first step are compared (using the same rules) to the best model from the first step. Finally, the best model from this step is compared to $\{\Omega\}$, to confirm that there is a difference in parameters or groups. This process stops when a more complex model is significantly different from every simpler nested model.

Each model requires starting values in a list of three objects where each object contains either one value for the parameters that do not differ among groups or as many values as groups considered for the parameters that do differ among groups in the model. In situations where similar starting values may be used for all groups, Map() and the starting values from $\{\Omega\}$ can be used to efficiently create starting values for all groups.[15] For example, the starting values for $\{\Omega\}$ (in sv0) are expanded to create starting values for $\{L_\infty, K\}$ as follows.

```
> ( svLK <- Map(rep,sv0,c(2,2,1)) )
$Linf
[1] 1193.483 1193.483

$K
[1] 0.2143151 0.2143151

$t0
[1] -1.121557
```

The list of starting values for the two other models to be compared to $\{L_\infty, K, t_0\}$ are created similarly.

```
> svLt <- Map(rep,sv0,c(2,1,2))
> svKt <- Map(rep,sv0,c(1,2,2))
```

The three nested models to be compared to $\{L_\infty, K, t_0\}$ are then fit and assigned to objects.

```
> fitLK <- nls(vbLK,data=bdmf,start=svLK)
> fitLt <- nls(vbLt,data=bdmf,start=svLt)
> fitKt <- nls(vbKt,data=bdmf,start=svKt)
```

The lrt() or extraSS() function is used to efficiently compare each of these nested models to $\{L_\infty, K, t_0\}$. In this case, the objects for all nested models are entered into lrt() or extraSS() as separate arguments and the object for the more complex $\{L_\infty, K, t_0\}$ model is entered in com=. The use of lrt() is illustrated here, though the process is the same for extraSS().

```
> lrt(fitLK,fitLt,fitKt,com=fitLKt,com.name="All pars diff",
      sim.names=c("Linf,K diff","Linf,t0 diff","K,t0 diff"))
Model 1: Linf,K diff
Model 2: Linf,t0 diff
Model 3: K,t0 diff
Model A: All pars diff

     Df0   logLik0 DfA   logLikA Df    logLik  Chisq Pr(>Chisq)
1vA  103 -561.3773 102 -561.2295  1   -0.1478 0.2956    0.58666
2vA  103 -561.2550 102 -561.2295  1   -0.0255 0.0510    0.82137
3vA  103 -563.1212 102 -561.2295  1   -1.8917 3.7834    0.05176
```

From this, there is strong evidence that $\{L_\infty, K\}$ ($p = 0.5867$) and $\{L_\infty, t_0\}$ ($p = 0.8214$) fit the data as well as $\{L_\infty, K, t_0\}$. The $\{L_\infty, t_0\}$ model is used as the basis for further model reduction in the next step because it has the greatest log-likelihood value (-561.255).

The starting values and model fits for the two most complex models that are nested subsets of $\{L_\infty, t_0\}$ (i.e., $\{L_\infty\}$ and $\{t_0\}$) and the likelihood ratio comparisons to $\{L_\infty, t_0\}$ are computed below.

```
> svL <- Map(rep,sv0,c(2,1,1))
> svt <- Map(rep,sv0,c(1,1,2))
> fitL <- nls(vbL,data=bdmf,start=svL)
> fitt <- nls(vbt,data=bdmf,start=svt)
> lrt(fitL,fitt,com=fitLt,com.name="Linf,t0 dif",
      sim.names=c("Linf dif","t0 dif"))
Model 1: Linf dif
Model 2: t0 dif
Model A: Linf,t0 dif

     Df0  logLik0 DfA  logLikA Df   logLik Chisq Pr(>Chisq)
1vA  104 -561.5209 103 -561.2550  1  -0.2660 0.5320    0.46577
2vA  104 -564.2286 103 -561.2550  1  -2.9737 5.9473    0.01474
```

These results suggest that $\{L_\infty\}$ ($p = 0.4658$), but not $\{t_0\}$ ($p = 0.0147$), fits as well as $\{L_\infty, t_0\}$. Thus, $\{L_\infty\}$ is the basis for the final step of the model reduction.

The comparison of $\{L_\infty\}$ to $\{\Omega\}$ confirms that a difference between groups exists and that that difference is only evident for L_∞ ($p = 0.0033$).

```
> lrt(fit0,com=fitL,com.name="Linf dif",sim.names="No pars dif")
Model 1: No pars dif
Model A: Linf dif

     Df0  logLik0 DfA  logLikA Df   logLik Chisq Pr(>Chisq)
1vA  105 -565.8258 104 -561.5209  1  -4.3049 8.6097   0.003344
```

Model Selection with AIC or BIC

The AIC or BIC can also be used to identify the "best" model from a suite of models fit to the same data (Burnham and Anderson 2002). Both measures were briefly described in Section 9.3.2. The model with the lowest AIC or BIC (depending on which measure you choose to use[16]) is considered the best fit to the data. The AIC and BIC values for a set of models are determined with AIC() or BIC(), respectively, with model objects as arguments.

The one model from Table 12.1 that was not fit in the sequential process above (i.e., $\{K\}$) is fit below to complete the fitting of all eight models.

```
> svK <- Map(rep,sv0,c(1,2,1))
> fitK <- nls(vbK,data=bdmf,start=svK)
```

All eight models are then given to AIC() and BIC() below, with the results column-bound together for a succinct presentation.

```
> cbind(AIC(fitLKt,fitLK,fitLt,fitKt,fitL,fitK,fitt,fitO),
+       BIC(fitLKt,fitLK,fitLt,fitKt,fitL,fitK,fitt,fitO))
        df      AIC df      BIC
fitLKt   7 1136.459  7 1155.234
fitLK    6 1134.755  6 1150.847
fitLt    6 1134.510  6 1150.603
fitKt    6 1138.242  6 1154.335
fitL     5 1133.042  5 1146.453
fitK     5 1136.424  5 1149.834
fitt     5 1138.457  5 1151.868
fitO     4 1139.652  4 1150.380
```

The L_∞ model has the lowest AIC (1133.042) and BIC (1146.453) and, thus, would be considered the "best" model using either criterion.

12.4.3   Summarizing the Model Fit

The comparison of multiple VBGF fits may be presented in one of two ways. First, the final model from the model reduction shown in Section 12.4.2 is presented as is (both groups are modeled with a common K and t_0, but different L_∞ values). Second, the results for each group are presented separately and the p-values from the model reduction are discussed to explain which parameters and groups differ significantly. The latter presentation is shown below.

The code below creates a function for the Typical VBGF (repeated from above), creates separate subsets of female and male Black Drum, and largely follows the methods from Section 12.3.3 to fit the model and extract parameter estimates and bootstrap confidence intervals for each sex.

```
> vbTyp <- vbFuns("typical")
> ## Females
> bdf <- filterD(bdmf,sex=="female")
> fitf <- nls(tl~vbTyp(otoage,Linf,K,t0),data=bdf,start=svO)
> bcf <- nlsBoot(fitf)
> cbind(coef(fitf),confint(bcf))
                      95% LCI      95% UCI
Linf 1226.9804659 1204.5011506 1249.3090080
K       0.1385006    0.1209127    0.1590749
t0     -1.9188267   -2.6900563   -1.2640084
```

```
> ## Males
> bdm <- filterD(bdmf,sex=="male")
> fitm <- nls(tl~vbTyp(otoage,Linf,K,t0),data=bdm,start=svO)
> bcm <- nlsBoot(fitm)
> cbind(coef=coef(fitm),confint(bcm))
              coef      95% LCI      95% UCI
Linf 1196.7191329 1179.4716889 1216.7118314
K       0.1418269    0.1238433    0.1624316
t0     -1.5943278   -2.5241348   -0.7975309
```

Constructing a visual of the separate fits (Figure 12.10) is more involved here than it was in Section 12.3.5 and, thus, is described in steps. First, predicted lengths across the range of ages are computed separately for males and females. The range of values for both ages (i.e., xlmts) and predicted lengths (i.e., ylmts) are also stored.

```
> # predictions for females
> xf <- seq(min(bdf$otoage),max(bdf$otoage),length.out=199)
> pf <- vbTyp(xf,Linf=coef(fitf))
> # predictions for males
> xm <- seq(min(bdm$otoage),max(bdm$otoage),length.out=199)
> pm <- vbTyp(xm,Linf=coef(fitm))
> xlmts <- range(c(xf,xm))
> ylmts <- range(c(bdf$tl,bdm$tl))
```

Second, a base plot is constructed with the plotting color set to white, the same as the background, so that no points are seen. The actual points are added next.

```
> plot(tl~otoage,data=bdmf,xlab="Age",ylab="Total Length (mm)",
       xlim=xlmts,ylim=ylmts,col="white")
```

Third, points for males and females are added to the base plot with different symbols (different semitransparent colors would show more contrast between the sexes).

```
> points(tl~otoage,data=bdf,pch=1,col=rgb(0,0,0,1/2),cex=0.8)
> points(tl~otoage,data=bdm,pch=8,col=rgb(0,0,0,1/2),cex=0.8)
```

Finally, the best-fit models and a legend are added to the plot.

```
> lines(pf~xf,lwd=2,lty="solid")
> lines(pm~xm,lwd=2,lty="dashed")
> legend("bottomright",c("Female","Male"),pch=c(1,8),
         lwd=2,lty=c("solid","dashed"),bty="n",cex=0.8)
```

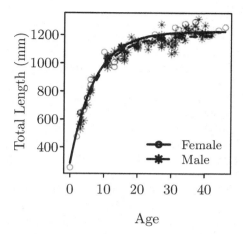

FIGURE 12.10. Length versus age with separate superimposed best-fit von Bertalanffy growth functions for male and female Black Drum.

12.5 Typical Model Fitting Problems

Fish growth data are notoriously difficult to fit with a VBGF. Problems may be due to the extent of the data, characteristics of the model, or both. Bolker et al. (2013) identified, explained, and addressed a number of general issues with fitting nonlinear models. Specific problems with fitting nonlinear growth models to fish are discussed here.

A common problem with growth data is the lack of either young or old fish. Young fish are often not present in the sample due to sampling difficulties (e.g., gear selectivity). The lack of old fish in the sample may be due to sampling a population where the mortality rate is high enough that fish do not live long enough to approach the asymptote (Francis 1988), difficulties capturing large old fish (e.g., gear selectivity), or from underestimating the age of older fish (i.e., older fish would appear younger). In any case, the lack of either of these groups of fish may restrict the ages to a narrow range that is functionally linear. It may be difficult to fit a nonlinear function, especially one with a horizontal asymptote, to relatively linear data.

One way to deal with the lack of young fish is to use a different version[17] of the VBGF,

$$E[L|t] = L_\infty - (L_\infty - L_0)\, e^{-Kt}$$

where t_0 is replaced with L_0, the mean length at time zero (von Bertalanffy 1938). When using this VBGF,[18] some authors (see citations in Cailliet et al. (2006)) have addressed the problem with young fish by fixing L_0 at a biologically derived value (e.g., length-at-hatch). This can, however, greatly impact the estimates of the other parameters (Cailliet et al. 2006).

Fish length-at-age data are also commonly quite variable (the Black Drum data are not a good example of this). This variability is likely largely due to ecological plasticity, but may also be due to a tendency to underestimate the age of older fish. Nevertheless, it may be difficult to fit a complex nonlinear function to highly dispersed data.

A primary difficulty with the Typical VBGF is that it has very highly correlated parameters. Parameters that are very highly correlated suggest that there is very little difference in model fit among many different combinations of the parameters. In other words, highly correlated parameters result in an RSS surface (e.g., Figure 12.2) that is "flat" near the global minimum. Thus, the gradient to this global minimum is shallow and the algorithms move slowly or erratically toward that global minimum, which makes the global minimum very difficult to find (Bolker et al. 2013).

Expected value parameterizations of models tend to have much less correlated parameters. Francis (1988) provided a version of the VBGF with all parameters defined as expected values

$$E[L|t] = L_1 + (L_3 - L_1) \frac{1 - r^{2\frac{t-t_1}{t_3-t_1}}}{1 - r^2}$$

with

$$r = \frac{L_3 - L_2}{L_2 - L_1}$$

where L_1, L_2, and L_3 are the mean lengths at ages t_1, t_2, and t_3, with t_1 and t_3 chosen by the analyst as relatively "young" and "old" ages, respectively, and $t_2 = \frac{t_1+t_3}{2}$. Thus, if issues arise with the Typical VBGF because of correlated parameters, then one may be able to fit the Francis VBGF to the data.[19] An example of fitting the Francis parameterization of the VBGF to the male Black Drum data is provided in the online supplement.

Modifications to the model may be helpful in some situations. The model is more likely to converge if there are fewer parameters. In some instances, as mentioned above, one of the parameters in the model may be set to a constant and, thus, only two other parameters need be estimated. It might also be useful if the parameter values are rescaled to be of roughly the same order of magnitude (Bolker et al. 2013). The problem here is usually related to L_∞; thus, for example, using lengths measured in meters or centimeters rather than millimeters may help with model convergence. In addition, rescaling may also be accomplished by specifying parameters on the lag scale, but entering them as exponentiated in the actual growth function (Bolker 2008). For example, this rescaling for L_∞ would result in

$$E[L|t] = e^{log_e(L_\infty)} \left(1 - e^{-K(t-t_0)}\right)$$

Modifications to `nls()` may also be helpful in some situations. The iterative fitting algorithm in `nls()` will stop, by default, after 50 iterations. The algorithm also works by halving the "step size" between iterations as the algorithm moves closer to the optimal solution. A minimum step size is set by default and if the working step size falls below this value before an optimal solution is reached, then the algorithm will terminate. The maximum number of iterations is increased and the minimum step size reduced by setting `maxiter=` and `minFactor=` in a list given to `control=` in the `nls()` call.

```
> fitTyp <- nls(tl~vbTyp(otoage,Linf,K,t0),data=bdm,start=svTyp,
               control=list(maxiter=100,minFactor=1/5000))
```

The default algorithm in `nls()` is a Gauss-Newton algorithm. In some situations, this algorithm will return an error about "infinity in predictions" or result in unreasonable parameter estimates (e.g., $K < 0$). These situations may be remedied by providing constraints on the parameter values to be considered during the fitting process. Using an exponentiated log-transformed parameter as described above effectively constrains that parameter to positive values. Alternatively, the "port" algorithm (include `algorithm="port"` to `nls()`) allows specific constraints for parameters to be defined. Lower and upper bounds for the parameters are included in lists given to `lower=` and `upper=` in `nls()`. The parameter constraints in each list must be in the same order as in the starting values list. Generally, the constraints should be set liberally (i.e., wide bounds) so that the algorithm does not often return an estimate from the boundary conditions.

```
> fitTypP <- nls(tl~vbTyp(otoage,Linf,K,t0),data=bdm,
                start=svTyp,algorithm="port",
                lower=list(Linf=1000,K=0.05,t0=-4),
                upper=list(Linf=1400,K=0.30,t0=1))
```

The algorithms in `nls()` may fail with a "singular gradient" error. This error is usually associated with poor choices for starting values. However, other algorithms have been implemented in R that may handle this issue better. For example, `nlsLM()` from **minpack.lm** (Elzhov et al. 2013) uses the Levenberg-Marquardt algorithm, `nlxb()` from **nlmrt** (Nash 2014) uses the Marquardt-Nash approach, and `nls2()` from **nls2** (Grothendieck 2013) implements a grid-search methodology. Use of these functions is illustrated in the online supplement.

12.6 Further Considerations

Other growth functions may be fit using methods similar to those described in this chapter. Different parameterizations of the VBGF are available in vbFuns() (see type= for vbFuns() and vbModels()). Other common growth functions available in **FSA** are the Gompertz with GompertzFuns(), logistic with logisticFuns(), Richards with RichardsFuns(), and Schnute with schnute(). Functions to model seasonal effects on growth (Pitcher and Mac-Donald 1973; Cloern and Nichols 1978; Pauly and Gaschutz 1979; Somers 1988; Pauly et al. 1992) and to work with tag-recapture data (Fabens 1965; Wang et al. 1995; Troynikov et al. 1998; Wang 1998; Haddon et al. 2008; Helidoniotis et al. 2011) also exist, several of which are available in vbFuns(), GompertzFuns(), or logisticFuns().

The following works use statistical methods to model fish growth that are beyond the introductory scope of this book. Kimura (2000) and Cope and Punt (2007) used nonlinear functional regressions and models with random effects, respectively, to incorporate aging error into the VBGF. Kimura (2008) provided a more general approach than that used in Section 12.4 to assess the effects of covariates, including a variable that identifies groups, on growth functions. Szalai et al. (2003) incorporated a time-varying random walk component into a VBGF to describe model parameters over time. Pilling et al. (2002), Schaalje et al. (2002), Vigliola and Meekan (2009), and Escati-Penaloza et al. (2010) used mixed effects models to analyze repeated measures data or to model individual variability in growth parameters. Recently, Bayesian models were used to analyze fish growth by He and Bence (2007), Shelton and Mangel (2012), Schofield et al. (2013), and Midway et al. (2015). Finally, Helser and Lai (2004) took a Bayesian hierarchical meta-analysis approach to summarize growth of Largemouth Bass at the continental scale.

Notes

[1] More thorough treatments of nonlinear models are found in Bates and Watts (2007) and Ritz and Streibig (2008).

[2] Data manipulations in this chapter require functions from **magrittr** and **dplyr**, which are fully described in Chapter 2.

[3] Vigliola and Meekan (2009) provided example R code for fitting a mixed model to length-at-age data.

[4] Examples of using other growth functions are in the online supplement.

[5] This use of dynamic plots requires the **relax**, **tcltk**, and **tkrplot** packages which may require manual installation in some environments. See Section 1.3 for instructions on how to manually install packages in R.

[6] The Typical parameterization of the VBGF can also be fit with growth() from **fish-**

methods. The Francis parameterization of the VBGF can also be fit with `vbfr()` of **fishmethods**.

[7]The actual profile likelihoods are seen with `plot()` and `profile()` from **MASS** (Venables and Ripley 2002). For example, `plot(profile(fitTyp),conf=0.95)`. The profile likelihood method for constructing confidence intervals is described generally in Hilborn and Mangel (1997), Bolker (2008), and Haddon (2011).

[8]The reader's results may be different than the results shown here because of the randomization used in the bootstrapping process. The reader may control the randomization by setting the random number seed with `set.seed()`. The reader can see the use of `set.seed()` in the script for this chapter.

[9]Monte Carlo simulations and Taylor approximations may also be used to construct confidence and prediction intervals for predicted lengths-at-age. These methods are implemented in `predictNLS()` of **propagate** (Spiess 2014).

[10]See Section 13.1.3.6 for a method to construct confidence polygons.

[11]Ritz and Streibig (2008) described methods to assess each assumption in greater detail. In addition, they provided more methods for handling assumption violations, including alternative methods for handling heteroscedasticity.

[12]The method of Kimura (1980) is most often cited for comparing VBGFs between groups. Kimura's methodology is the same as that described in Section 12.4.2 when `lrt()` is used, with the exception that Kimura stopped after comparing the most complex nested submodels of $\{L_\infty, K, t_0\}$ to $\{L_\infty, K, t_0\}$. The methods described in Section 12.4.1 are more general. Kimura's less general method is also implemented in `vblrt()` of **fishmethods**.

[13]These naming conventions follow that of Ritz and Streibig (2008).

[14]In instances where the groups do differ dramatically, the starting values are manually entered into a named list. For example, with only two groups (i.e., sexes), the starting values might be entered as `svLKt <- list(Linf=c(1200,1000),K=c(0.13,0.31),t0=c(-2,0))`.

[15]If the groups differ dramatically, then the list of starting values must be created manually. For example, the list of starting values for $\{L_\infty, K\}$ might be entered as `svLK <- list(Linf=c(1200,1000),K=c(0.13,0.31),t0=-2)`.

[16]There is considerable confusion and debate about whether to use the AIC or BIC metric. The interested reader could begin an exploration of this topic with Burnham and Anderson (2004) or Aho et al. (2014).

[17]Models can often be rewritten to have different parameters. These different versions are called *parameterizations*. Different parameterizations of the same model fit the data exactly the same way (i.e., predicted values of the response (or dependent) variable are equivalent) and have the same number of parameters (i.e., they are not more parsimonious). Different parameterizations are used to estimate different parameters, though there may also be algorithmic considerations (see further in this section).

[18]In **FSA**, this VBGF is declared with `type="Original"` in `vbStarts()` and `vbFuns()`.

[19]In **FSA**, this model is declared with `type="Francis"` in `vbStarts()` and `vbFuns()`.

13

Recruitment

At its most basic level, a "recruit" is a "young" fish that is added to the population. However, how "young" is defined may depend on context (Allen and Hightower 2010; Haddon 2011). For example, fish that have matured to an *age* where they are able to reproduce have "recruited to the breeding population" and fish that have survived to age-1 have "recruited to age-1." Moreover, fish that have survived to a *length* where they are able to reproduce have "recruited to the breeding population" and fish that have survived to a catchable size (defined by tackle, gear, or minimum size by regulation) have "recruited to the fishery." Thus, a recruit may be defined by a critical period in the fish's life history that is defined by either the length or age of the fish. Most stock-recruit analyses will consider "recruitment to the fishery" based on length or "recruitment to the breeding population" based on length or age.

Recruitment is defined as the number of fish that survive to the length or age to be called a recruit. Identifying the number of recruits, either absolutely or relatively, is important to fisheries scientists because this value gives insight into the future viability of or possible harvest from the population. In this chapter, models that examine the relationship between the number of reproducing fish and recruitment, number of reproducing females and egg production, and indices of year-class or cohort strength are examined.

Required Packages for This Chapter

Functions used in this chapter require loading the packages shown below.[1]

```
> library(FSA)
> library(car)        # Before dplyr to reduce conflicts with MASS
> library(dplyr)
> library(magrittr)
> library(plotrix)
> library(nlstools)
> library(lsmeans)
```

13.1   Stock-Recruitment Relationships

13.1.1   Data Requirements

Stock-recruit data consists of estimates of the number (or biomass) of reproducing fish and the number (or biomass) of subsequent recruits. The reproducing fish may be called "adults" or "spawners" and the number of spawners is called "spawning stock" or "stock" for short. The number of subsequent recruits is called "recruitment." Hereafter, stock and recruitment variables will be abbreviated with S and R, respectively.

Stock-recruitment data are notoriously messy with high year-to-year variability in R and weak relationships between R and S (Figure 13.1). This variability, or the resultant difficulty in fitting functions to these data, has been commented on by several authors (e.g., Quinn II and Deriso 1999; Hilborn and Walters 2001). Wootton (1998, p. 241) was particularly blunt when he said

> Much ingenuity has been spent in fitting these curves to data sets and to elaborating the basic models. All this effort has largely foundered in the face of the variability in the relationships between stock and recruitment shown by most natural populations. The curves can be fitted, but it takes an act of faith to take the resulting curves seriously.

Despite this, assessing the relationship between R and S is important because, as Hilborn and Walters (2001, p. 241) stated

> Fisheries managers simply cannot ignore the fact that if you fish hard enough on any stock, you will reduce recruitment. While recruitment may be largely independent of stock size as a fishery develops, experience has shown that most fisheries will reach the point where recruitment begins to drop due to overfishing. The problem for biologists is to try to understand the relationship between stock and recruitment at least well enough to know how much the stock can be reduced before recruitment starts to drop.

Subbey et al. (2014, p. 2318) extended this into the future when they wrote

> There is not doubt that, in an ever changing climate, recruitment modelling and forecasting will remain central to fisheries science in the next 100 years.

The methods of this section are intended to help understand the relationship between R and S.

Data on the escapement (`esc`, in thousands of fish; i.e., the stock) and returns two years later (`ret`, in thousands of fish; i.e., recruitment) of Pink

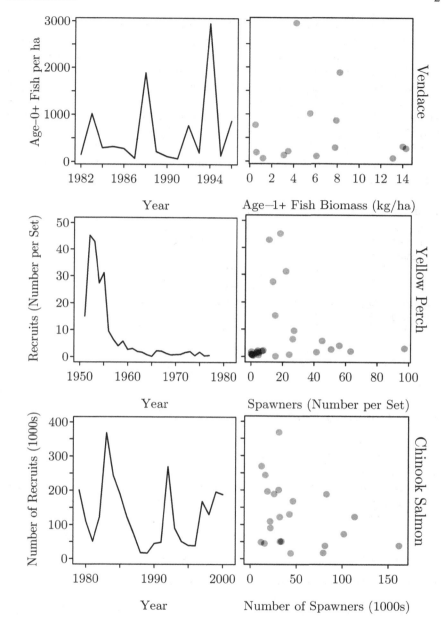

FIGURE 13.1. Plot of number of recruits versus year (Left) and versus number of spawners (Right) for three example species. The Vendace data (top row) are from Lake Puulavesi (Finland) from 1982–1996 (Marjomaki 2004). The Yellow Perch data (middle row) are from South Bay, Lake Huron (Ontario) from 1950–1983 (Henderson 1985). The Chinook Salmon data (bottom row) are from the 1979–2000 brood years for the Klamath River (California; from the Pacific Fishery Management Council).

Salmon in southeast Alaska recorded from 1960 to 1991 (from Quinn II and Deriso 1999) are used in this section to illustrate the fitting of stock-recruitment functions. In addition to the escapement and return variables, the sea surface temperature (SST) one year later (when Pink Salmon were in the ocean) was recorded. Because of the time lags in the data (returns were measured two years later), missing values occur in the first and last two rows of the data.frame. These rows were removed below. In addition, for simplicity in plotting, both escapement and returns were divided by 1000 so that those variables now record millions of fish (a thousand thousands). Finally, for purposes described later, the natural logarithm of the returns was also created (with `log()`).

```
> pinks <- read.csv("PSalmonAK.csv") %>%
    filter(!is.na(esc),!is.na(ret)) %>%
    mutate(esc=esc/1000,ret=ret/1000,logret=log(ret))
> headtail(pinks)
   year   esc    ret   SST   logret
1  1960 1.418  2.446 11.63 0.894454
2  1961 2.835 14.934 11.46 2.703640
3  1962 1.957 10.031 12.33 2.305680
28 1987 4.289 18.215 11.81 2.902245
29 1988 2.892  9.461 12.26 2.247178
30 1989 4.577 23.359 12.30 3.150982
```

13.1.2 Common Models

There are many functions used to model the relationship between S and R (Needle 2002; Maceina and Pereira 2007; Subbey et al. 2014). However, three of these functions appear to be the most common.

The *Beverton-Holt* function (Beverton and Holt 1957) assumes that R will approach an asymptote at high values of S (Figure 13.2).[2] This function has a variety of parameterizations[3] including

$$E[R|S] = \frac{aS}{1+bS} \tag{13.1}$$

where $E[R|S]$ is the expected (i.e., mean) R for a given value of S, a is often called the *density-independent* parameter, and b is the *density-dependent* parameter (Quinn II and Deriso 1999; Maceina and Pereira 2007). The asymptote, or peak R, for the Beverton-Holt function is $R_p = \frac{a}{b}$.

The *Ricker* function (Ricker 1954) assumes a dome-shaped relationship where the peak R occurs at an intermediate S (Figure 13.2).[4] This function also has a variety of parameterizations including

$$E[R|S] = aSe^{-bS} \tag{13.2}$$

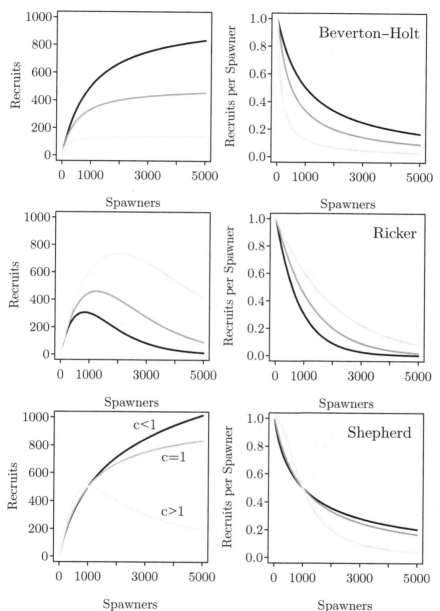

FIGURE 13.2. Idealistic plots of number of recruits (Left column) and number of recruits per spawner (Right column) against number of spawners for the Beverton-Holt (top row), Ricker (middle row), and Shepherd (bottom row) stock-recruitment functions. All functions use $a = 1$ and varying values of b for the Beverton-Holt and Ricker functions, and $b = 0.001$ and varying values of c for the Shepherd function. Line colors are darker for smaller values of the varying parameter.

where a and b are again the density-independent and density-dependent parameters (Quinn II and Deriso 1999). The peak R for the Ricker function is $R_p = \frac{a}{be}$ and occurs at $S_p = \frac{1}{b}$.

The *Shepherd* function (Shepherd 1982) includes a third parameter (c) that allows for a variety of shapes; that is,

$$E[R|S] = \frac{aS}{1 + (bS)^c} \tag{13.3}$$

If $c = 1$, then the Shepherd function reduces to the Beverton-Holt function and if $c > 1$, then the shape of the Shepherd function resembles that of the Ricker function (Figure 13.2).

Finally, if density dependence does not exist in these stock-recruitment relationships, then $b = 0$ and Equations 13.1, 13.2, and 13.3 reduce to a linear function through the origin with a constant slope of a; that is,

$$E[R|S] = aS \tag{13.4}$$

This function is called the *density-independent* stock-recruitment function and is used in later sections to statistically test the significance of the density-dependent parameter (b) in the Beverton-Holt, Ricker, and Shepherd functions. In addition, it can be seen from dividing both sides of Equation (13.4) by S that the units of a are "recruits per spawner."

13.1.3   Fitting Models

The Beverton-Holt, Ricker, and Shepherd stock-recruitment functions are nonlinear. Thus, the background, methodology, cautions, and code described for growth functions in Section 12.3 are pertinent, but will not be repeated, here. Consult Section 12.3 for specific details on fitting nonlinear regressions.

13.1.3.1   Starting Values

Starting values (denoted with a ~ over the parameter) for stock-recruitment functions may be obtained from parameter estimates from linearized versions of the functions or from a visual fit of the function to data. For completeness, both methods are illustrated below.

The linearized form[5] of the Beverton-Holt function in Equation (13.1) is

$$\frac{1}{E[R|S]} = \frac{1}{a}\frac{1}{S} + \frac{b}{a}$$

where the slope is $\frac{1}{a}$ and the intercept is $\frac{b}{a}$. Manipulation of these relations gives $\tilde{a} = \frac{1}{slope}$ and $\tilde{b} = \frac{intercept}{slope}$, where the *intercept* and *slope* are estimated from a linear regression of $\frac{1}{R}$ on $\frac{1}{S}$.

The linearized form[6] of the Ricker function in Equation (13.2) is

$$log\left(\frac{E[R|S]}{S}\right) = log(a) - bS \qquad (13.5)$$

where the slope is $-b$ and the intercept is $log(a)$. Manipulation of these relations gives $\tilde{a} = e^{intercept}$ and $\tilde{b} = -slope$, where the *intercept* and *slope* are estimated from a linear regression of $log(\frac{R}{S})$ on S.

Starting values for a and b in the Shepherd function may be the same as the Beverton-Holt function. The starting value for c can likely be 1. However, if the stock-recruitment relationship has a strong "dome-shape," then a starting value for c greater than 1 will likely perform better.

These methods for deriving starting values are implemented in `srStarts()` of **FSA** with a formula of the form R~S in the first argument, where R and S generically represent the R and S variables in the data.frame in `data=`. A name for the stock-recruitment function to be used is given in `type=` (i.e., "BevertonHolt", "Ricker", or "Shepherd").[7]

```
> ( svR <- srStarts(ret~esc,data=pinks,type="Ricker") )
$a
[1] 2.849252

$b
[1] 0.05516673
```

Starting values for the stock-recruitment function may also be obtained by extracting parameters from a visual fit of the function to the observed data. A scatterplot with the stock-recruitment function superimposed on the data (Figure 13.3), but with parameters connected to slider bars that adjust the superimposed function when moved, is constructed by including `dynamicPlot=TRUE` in `srStarts()`.[8] The parameters may be altered with the slider bars until an approximate fit is obtained. The slider bar values at that point are then entered into a named list to serve later as starting values.

```
> srStarts(ret~esc,data=pinks,type="Ricker",dynamicPlot=TRUE)
> svR <- list(a=2.8,b=0.6)
```

13.1.3.2 Error Structure

Quinn II and Deriso (1999) and Hilborn and Walters (2001) suggested that multiplicative, rather than additive, errors should be the default choice when fitting stock-recruitment functions.[9] Subbey et al. (2014) noted that multiplicative errors are commonly assumed for stock-recruitment models. Thus, multiplicative errors will be assumed when fitting stock-recruitment functions here (i.e., logarithms will be used as described in Section 12.3.6). Regardless, one should carefully examine the residuals of the fitted model before settling on an error structure.[10]

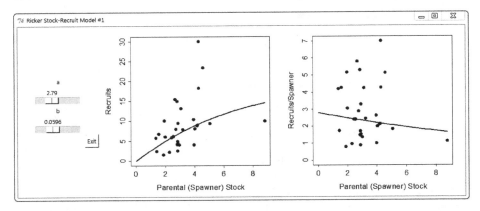

FIGURE 13.3. Example of using dynamic plots (e.g., `dynamicPlot=TRUE` in `srStarts()`) to find an approximate fit of the Ricker stock-recruitment function for Pink Salmon.

13.1.3.3 Parameter Estimates and Confidence Intervals

For efficiency, the stock-recruitment function should be entered into an R function. R functions for the three stock-recruitment functions mentioned in this chapter may be created with `srFuns()` from **FSA** with the appropriate name (i.e., `"BevertonHolt"`, `"Ricker"`, or `"Shepherd"`) as the only argument.[11] The result of `srFuns()` is assigned to an object which then becomes the name of the created function. The function created by `srFuns()` allows values for all parameters to be given in the first argument, which will be useful later. The code below creates an R function for the Ricker function and demonstrates how it can be used to predict a value of R given $S = 2$ (i.e., an escapement of 2 million) and values for the parameters.

```
> rckr <- srFuns("Ricker")
> rckr(S=2,a=2.8,b=0.6)        # parameters separate
[1] 1.686688
> rckr(S=2,a=c(2.8,0.6))       # parameters in one vector
[1] 1.686688
```

The stock-recruitment function with multiplicative errors is fit with `nls()` using a formula that has the log recruitment variable on the left-hand side and the log of the stock-recruit function created above on the right-hand side.

```
> srR <- nls(logret~log(rckr(esc,a,b)),data=pinks,start=svR)
```

Parameter estimates and profile likelihood confidence intervals (CI) are extracted from the `nls()` object with `coef()` and `confint()`, respectively (column-bound below for display).

```
> cbind(estimates=coef(srR),confint(srR))
```
Waiting for profiling to be done...
```
    estimates        2.5%       97.5%
a 2.84925231   1.6127074 5.0365567
b 0.05516673  -0.1065979 0.2169314
```

Confidence intervals for the parameter estimates may also be derived via bootstrapping with `nlsBoot()` from **nlstools** followed by `confint()`.

```
> bootR <- nlsBoot(srR)
> cbind(estimates=coef(srR),confint(bootR))
    estimates       95% LCI     95% UCI
a 2.84925231   1.72579680 4.8562503
b 0.05516673  -0.09376845 0.2072737
```

One or more "fit did not converge" warnings may be issued by `nlsBoot()`. These warnings are fairly common when fitting stock-recruitment functions because of the often weak relationship between R and S. You should question the model fit if the number of convergence warnings exceeds 10.

13.1.3.4   Model Predictions

The R can be predicted for a value of S using the function created above (i.e., `rckr()`) and the coefficients from the fitted model. For example, the predicted mean R for $S = 2$ (i.e., an escapement of 2 million fish) using the fitted model is computed below.

```
> rckr(S=2,a=coef(srR))
[1] 5.103213
```

A confidence interval for the predicted mean R is obtained by predicting the mean R for each bootstrap sample using `apply()` and then extracting the 2.5 and 97.5 percentile of those results with `quantile()`.

```
> pR <- apply(bootR$coefboot,MARGIN=1,FUN=rckr,S=2)
> quantile(pR,c(0.025,0.975))
    2.5%      97.5%
3.864964 6.682341
```

Thus, an escapement of 2 million fish is predicted to produce a mean return of 5.1 million fish (95% CI: 3.9, 6.7 million fish).

13.1.3.5   Assessing Parameter Significance

The statistical significance of the density-dependent parameter (b) is assessed by comparing the fits of Equation (13.2) and Equation (13.4). Equation (13.2) was fit above. Equation (13.4) is fit to the Pink Salmon data below.[12]

```
> ind <- srFuns("independence")
> svI <- srStarts(ret~esc,data=pinks,type="independence")
> srI <- nls(logret~log(ind(esc,a)),data=pinks,start=svI)
```

The models are then compared statistically with an extra sum-of-squares test (using `extraSS()`) or likelihood ratio test (using `lrt()`) as described in Section 12.4.2. Only the extra sum-of-squares test is shown below.

```
> extraSS(srI,com=srR)
Model 1: logret ~ log(ind(esc, a))
Model A: logret ~ log(rckr(esc, a, b))

    DfO     RSSO DfA     RSSA Df        SS     F Pr(>F)
1vA  29 10.62735  28 10.44530  1   0.18205 0.488 0.4906
```

The p-value ($p = 0.4906$) is large which suggests that the simpler independence model that does not include the b parameter fits these data adequately. Thus, it appears that escapement of Pink Salmon does not contribute significantly to explaining the variability in the returns of Pink Salmon two years later.

The traditional *coefficient of determination* (r^2) does not exist for nonlinear models. However, Maceina and Pereira (2007) described a *quasi-r^2* value that is the square of the Pearson correlation coefficient (r) computed (with `cor()`) between the observed and predicted values of R.

```
> cor(rckr(pinks$esc,a=coef(srR)),pinks$ret)^2
[1] 0.1840478
```

The exact interpretation of this quasi-r^2 is unknown but it is clear that low values, as in this example, indicate that little of the variability in R is explained by knowing S.

13.1.3.6 Visualizing the Model Fit

A visual of the fitted function with a 95% confidence band (Figure 13.4) is constructed largely as described in Section 12.3.4. The construction of this plot begins by predicting the mean R at many values of S and then using a loop to compute and extract the corresponding lower and upper confidence values for the mean R at each S from the bootstrap results.

```
> x <- seq(0,9,length.out=199)        # many S for prediction
> pR <- rckr(x,a=coef(srR))           # predicted mean R
> LCI <- UCI <- numeric(length(x))
```

```
> for(i in 1:length(x)) {              # CIs for mean R @ each S
    tmp <- apply(bootR$coefboot,MARGIN=1,FUN=rckr,S=x[i])
    LCI[i] <- quantile(tmp,0.025)
    UCI[i] <- quantile(tmp,0.975)
  }
> ylmts <- range(c(pR,LCI,UCI,pinks$ret))
> xlmts <- range(c(x,pinks$esc))
```

These results are then plotted. The code below, however, differs from Section 12.3.4 by displaying the 95% confidence band as a polygon rather than with two lines. The steps for this construction are to create a schematic plot first (i.e., use `col="white"` in `plot()`), add the polygon for the confidence band, add the points, and add the best-fit model. The polygon is added with `polygon()` which takes vectors of "x" and "y" values as the first two arguments. The `rev()` function is used to reverse the order of the object in its argument. Thus, the first two arguments to `polygon()` below plot the lower boundary of the polygon from left to right (i.e., the LCI values against x) followed by the upper boundary of the polygon from right to left (i.e., the `rev(UCI)` values against `rev(x)`).

```
> plot(ret~esc,data=pinks,xlim=xlmts,ylim=ylmts,col="white",
       ylab="Returners (millions)",
       xlab="Escapement (millions)")
> polygon(c(x,rev(x)),c(LCI,rev(UCI)),col="gray80",border=NA)
> points(ret~esc,data=pinks,pch=19,col=rgb(0,0,0,1/2))
> lines(pR~x,lwd=2)
```

13.1.4 Additional Explanatory Variables

Several authors have demonstrated (e.g., Quinn II and Deriso 1999; Maceina and Pereira 2007; Haddon 2011) or discussed (Subbey et al. 2014) how to add additional explanatory variables to a stock-recruitment function to explain additional variability in recruitment. Most of these authors have also suggested considerable caution when using this approach. Nevertheless, this approach is illustrated below for Pink Salmon by adding the annual sea surface temperature data to the Ricker stock-recruitment function.

The Ricker function in Equation (13.2) is modified for an additional explanatory variable X as

$$E[R|S] = aSe^{-bS+cX} \tag{13.6}$$

Additional variables may be added similarly.

As previously, this new stock-recruit function should be entered into an R function. However, `function()`, as described in Section 1.8.2, must be used

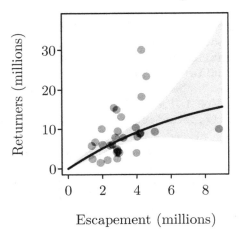

Escapement (millions)

FIGURE 13.4. Number of returners versus escapement for Pink Salmon in Alaska with the best-fit Ricker stock-recruitment function and 95% confidence polygon superimposed.

here because `srFuns()` cannot be flexible enough to handle all possible additional variables. Such a function is defined below (note that `if()` is required to allow all parameter values to be given in the first argument and `exp()` raises its argument to the power of e).

```
> rckr2 <- function(S,X,a,b=NULL,c=NULL) {
    if (length(a)>1) { # all values in a argument
      c <- a[3]
      b <- a[2]
      a <- a[1]
    }
    a*S*exp(-b*S+c*X)
  }
>
> rckr2(2,12,c(0.003,0.09,0.6)) # Example calculation
[1] 6.71272
```

Additionally, `srStarts()` cannot be used to identify starting values for Equation (13.6). A simple method to derive starting values for these extended stock-recruit functions is to estimate the parameters for the linearized version of the function (similar to Equation (13.5)) with regression methods (see Section 7.3).[13]

```
> tmp <- lm(log(ret/esc)~esc+SST,data=pinks)
> tmp <- coef(tmp)
> svR2 <- list(a=exp(tmp[[1]]),b=-tmp[[2]],c=tmp[[3]])
```

Now, with an appropriate R function and starting values for all parameters, the stock-recruitment function is fit to the data with `nls()`.

```
> srR2 <- nls(logret~log(rckr2(esc,SST,a,b,c)),
              data=pinks,start=svR2)
```

The extra SS test that simultaneously compares Equation (13.6) to Equation (13.2) and Equation (13.4) shows that the model with escapement **and** sea surface temperatures is preferred, with the model explaining slightly less than half of the variability in Pink Salmon returns.

```
> extraSS(srI,srR,com=srR2)
Model 1: logret ~ log(ind(esc, a))
Model 2: logret ~ log(rckr(esc, a, b))
Model A: logret ~ log(rckr2(esc, SST, a, b, c))

    DfO   RSSO DfA   RSSA Df      SS      F    Pr(>F)
1vA  29 10.6273  27 6.0578  2  4.5696 10.184 0.0005064
2vA  28 10.4453  27 6.0578  1  4.3875 19.556 0.0001437
> cor(rckr2(pinks$esc,pinks$SST,a=coef(srR2)),pinks$ret)^2
[1] 0.4527386
```

An examination of the model coefficients shows that the number of returning Pink Salmon increases with increasing sea surface temperature if escapement is held constant (i.e., the c coefficient is positive).

```
> coef(srR2)
          a           b           c
0.002216035 0.090309546 0.637709952
```

13.2 Spawning Potential Ratio

13.2.1 Single Values

Recruitment overfishing occurs when persistent fishing results in reduced yields due to a lack of recruits to the fishery (Maceina and Pereira 2007; Haddon 2011). Recruitment overfishing is likely more common in commercial than recreational fisheries (Allen and Hightower 2010). The typical stock-recruitment functions from the previous section have not been particularly useful for identifying recruitment overfishing (Maceina and Pereira 2007). An alternative metric with the potential to identify recruitment overfishing is the *spawning potential ratio* (SPR) developed by Goodyear (1989, 1993).

Goodyear (1993, p. 68) defined the *potential recruit fecundity* (P) as "the number of eggs that could be produced by an average recruit in the absence of density dependence." Maceina and Pereira (2007, p. 177) noted that P represents the "average lifetime production of mature eggs per recruit at the equilibrium population densities in the absence of any density-dependent suppression of maturation or fecundity at age." Potential recruit fecundity may be computed as

$$P = \sum_{i=1}^{n} \mu_i E_i S_i \tag{13.7}$$

where n is the maximum age, μ_i is the proportion of females that are mature at age i, E_i is the mean fecundity (number of eggs produced) in the absence of density-dependent effects on growth, maturity, or fecundity for *mature* females[14] of age i, and S_i is the cumulative survival rate (i.e., survivorship) of females from hatch to age i. The S_i are computed as

$$S_i = \prod_{j=0}^{i-1} S_{ij} \tag{13.8}$$

where S_{ij} is the annual survival rate of age i females when they were age j (for $j < i$). The S_{ij} are equal to $e^{-(F_{ij}+M_{ij})}$ where F_{ij} is the instantaneous fishing mortality rate on age i females when they were age j and M_{ij} is the instantaneous natural mortality rate of age i females when they were age j (see Section 11.1 for a discussion of instantaneous mortality rates).

Calculation of a single value of P is mostly a matter of bookkeeping and, if the S_{ij} are assumed to be constant across all i (e.g., the survival from age 3 to age 4 is the same for age-7 and age-8 fish), then the bookkeeping becomes rather simple. The calculation of P with this simplifying assumption is illustrated below.

The calculation of a single P is illustrated for a hypothetical population of female fish with a maximum age of 10 (`max_age`), a general increase in percent of fish that are mature to 100% maturity at age-4 (`pr_mat`), and a given schedule of mean fecundities-at-age (thousands of eggs; `fec`).

```
> max_age <- 10
> pr_mat <- c(0,0.25,0.75,1,1,1,1,1,1,1)
> fec <- c(0,31,53,106,160,213,266,319,373,426)
```

Furthermore, it is assumed that the population is affected by a constant $M = 0.3$ and, in this scenario, a constant $F = 0$ (i.e., no fishing mortality). Note that the `rep()` function is used below to create a vector where the value in the first argument is repeated the number of times given in the second argument.[15]

```
> M <- 0.3
> n_mort <- rep(M,max_age)
> F <- 0
> f_mort <- rep(0,max_age)
```

Under the simplifying assumption of consistent S_{ij} for all i, the annual survival rates are computed by summing the fishing and natural mortality rates and raising the negative of the sum to the power of e with exp().

```
> s_ann <- exp(-(n_mort+f_mort))
```

The survival rates to age i (i.e., Equation (13.8)) are then computed as a cumulative product using cumprod().

```
> s_cum <- cumprod(s_ann)
```

The parts on the right-hand side of Equation (13.7) are multiplied together.

```
> P_noF_i <- pr_mat*fec*s_cum
```

Each of the values computed above are column-bound together below with cbind() to provide a table of values at each age used in the calculation of P. The values are rounded to three significant digits below (using round()) for display purposes only.

```
> round(cbind(age=1:max_age,pr_mat,fec,n_mort,f_mort,
              s_ann,s_cum,P_noF_i),3)
       age pr_mat fec n_mort f_mort s_ann s_cum P_noF_i
 [1,]    1   0.00   0    0.3      0 0.741 0.741   0.000
 [2,]    2   0.25  31    0.3      0 0.741 0.549   4.253
 [3,]    3   0.75  53    0.3      0 0.741 0.407  16.161
 [4,]    4   1.00 106    0.3      0 0.741 0.301  31.927
 [5,]    5   1.00 160    0.3      0 0.741 0.223  35.701
 [6,]    6   1.00 213    0.3      0 0.741 0.165  35.209
 [7,]    7   1.00 266    0.3      0 0.741 0.122  32.573
 [8,]    8   1.00 319    0.3      0 0.741 0.091  28.939
 [9,]    9   1.00 373    0.3      0 0.741 0.067  25.068
[10,]   10   1.00 426    0.3      0 0.741 0.050  21.209
```

Finally, the right-hand side parts of Equation (13.7) (i.e., in P_noF_i) are summed to compute P.

```
> ( P_noF <- sum(P_noF_i) )
[1] 231.0399
```

Thus, under the assumptions of no fishing mortality and no density-dependent responses that affect fecundity, growth, or survival, one would expect that an average female in this population would produce 231.0 thousand eggs during her lifetime.

Goodyear (1993) proposed P primarily as an intermediate step to computing SPR, which is the ratio of P for a fished condition (P_{fished}) to P for an unfished condition ($P_{unfished}$); that is,

$$SPR = \frac{P_{fished}}{P_{unfished}}$$

The primary purpose of the SPR is to determine how lifetime egg production is affected under different management scenarios that affect F. For example, the population modeled above might be modified for some $F > 0$. Including an $F > 0$ is slightly more difficult because many fisheries have young fish that are not vulnerable, old fish that are fully vulnerable, and fish of intermediate age that are not fully vulnerable to the fishery.[16] This pattern of F may be modeled by creating a vector of "selectivities" that represents the proportions of maximum F that affect each age. For example, the following code models a fishery with $F = 0.6$ that does not affect the first two ages, affects half of the age-3 fish, and fully affects all fish age-4 and older.

```
> F <- 0.6
> sel <- c(0,0,0.5,1,1,1,1,1,1,1)
> ( f_mort <- F*sel )
 [1] 0.0 0.0 0.3 0.6 0.6 0.6 0.6 0.6 0.6 0.6
```

The survival rates are updated with these new fishing mortality rates so that P under this hypothetical fishery is computed.

```
> s_ann <- exp(-(n_mort+f_mort))
> s_cum <- cumprod(s_ann)
> ( P_F <- sum(pr_mat*fec*s_cum) )
[1] 45.48311
```

Thus, under this scenario of fishing mortality, one would expect that an average female in this population would produce 45.5 thousand eggs during her lifetime. Thus, the SPR would be $\frac{45.5}{231.0} = 0.197$. In other words, 19.7% of eggs that would be produced with no fishing morality would be produced under the $F = 0.6$ scenario.

13.2.2 Multiple Values

Several authors have suggested threshold values of SPR at which lower values would indicate potential recruitment overfishing. However, Mace and Sissenwine (1993) concluded that no one value of SPR could be set as a threshold for recruitment overfishing because SPR values are related to species- and population-specific parameters such as rates of natural mortality, maximum body weight, and size at 50% maturity. Nevertheless, Goodyear (1993) and Mace and Sissenwine (1993) suggested that SPR values greater than 0.20 or 0.30 would limit recruitment overfishing for marine pelagic species. In specific freshwater fisheries, the recommended SPR threshold is 0.10 for Channel Catfish in the Upper Mississippi River (Slipke et al. 2002) and 0.40 or 0.50 for Shovelnose Sturgeon in the Missouri River (Quist et al. 2002).

Spawning potential ratios may be computed under various levels of F (ideally tied to various management regulation scenarios) to identify a level of F that meets the threshold SPR value. Simulations of SPR are facilitated by wrapping the computations from above into a function that returns a SPR value when given values of M and F as well as vectors of E_i, μ_i, and capture selectivities as arguments. Such a function is created (using `function()` as described in Section 1.8.2) and checked below.

```
> SPR <- function(M,F,E,mu,sel) {
    # vector of mortality rates
    M <- rep(M,length(E))
    # P for the fished and unfished scenarioes
    P_fish <- sum(mu*E*cumprod(exp(-(M+sel*F))))
    P_unfish <- sum(mu*E*cumprod(exp(-M)))
    # Compute and return SPR value
    P_fish/P_unfish
  }
>
> # check function with same values used in previous section
> SPR(M=0.3,F=0.6,E=fec,mu=pr_mat,sel=sel)
[1] 0.1968626
```

The utility of this function is that it can be used to efficiently compute values of SPR for a range of M and F values. An illustration of this utility begins by creating separate vectors of M and F values from 0.01 to 1.0 in steps of 0.01.

```
> Ms <- seq(0.01,1,0.01)
> Fs <- seq(0.01,1,0.01)
```

The SPR value computed at each combined value of M and F will be stored in a two-dimensional matrix with values of M defining rows and values of F defining columns. Such a matrix is initially filled with NAs below using `matrix()`.

```
> SPRs <- matrix(nrow=length(Ms),ncol=length(Fs),
                 dimnames=list(Ms,Fs))
```

A double loop (see Section 1.9.1) is used to compute SPR values at the combined values of M and F with the function defined above. The first (i) loop selects a value of M for use in the inner (j) loop which selects values of F. Once the values of F are exhausted, the inner loop is complete, another value of M is selected in the outer loop, and the inner loop will cycle through all values of F for that value of M. This process is repeated until all values of M are exhausted. At each iteration of the loop, the SPR for the given values of M and F is calculated and stored in the appropriate cell of the SPRs matrix.

```
> for(i in 1:length(Ms))
    for(j in 1:length(Fs))
      SPRs[i,j] <- SPR(M=Ms[i],F=Fs[j],E=fec,mu=pr_mat,sel=sel)
```

The final result of this double loop is a matrix that contains the SPR computed for each chosen value of M (by row) and F (by column). This matrix is too large to show here because the step value for the M and F vectors was small. However, a portion of the upper-left corner of this matrix is shown below.

```
> SPRs[1:5,1:5]
          0.01      0.02      0.03      0.04      0.05
0.01 0.9500788 0.9030054 0.8586062 0.8167184 0.7771898
0.02 0.9504543 0.9037232 0.8596355 0.8180308 0.7787589
0.03 0.9508341 0.9044493 0.8606768 0.8193587 0.7803469
0.04 0.9512179 0.9051833 0.8617298 0.8207018 0.7819534
0.05 0.9516058 0.9059251 0.8627942 0.8220597 0.7835780
```

For example, the SPR for $M = 0.02$ and $F = 0.04$ is 0.818.

The results in the SPRs matrix may be viewed as a *contour plot* where M, F points with similar SPR values are connected by "contour" lines (Figure 13.5). This plot is constructed with contour() where the first two arguments are the vectors of M (x-coordinate) and F (y-coordinate) values and the third argument is the matrix of SPR values that correspond to the M and F sequences. The SPR values for each contour line are set with levels=. The size of labels for each contour line are controlled with labcex= (the default is 0.6) and the placement of these labels is controlled with method= (the default is to place the labels at the flattest point within the contour line, whereas method="edge" will place the labels on the edges of the contour lines). Other arguments used below were defined in Chapter 3.

```
> contour(x=Ms,y=Fs,z=SPRs,levels=seq(0,1,0.1),lwd=2,
          method="edge",labcex=0.7,
          xlim=c(0,1),ylim=c(0,1),xlab="Natural Mortality (M)",
          ylab="Fishing Mortality (F)")
```

The contour plot in Figure 13.5 may be used to identify approximate values of F, given a value of M, that will meet a threshold SPR value. For example, if $M = 0.3$, then the fisheries scientist might augment the contour plot with a vertical line at 0.3. Therefore, if the threshold SPR is 0.4, then management regulations should be enacted to hold F less than approximately 0.25.

```
> abline(v=0.3,lwd=2,lty="dashed",col="gray50")
```

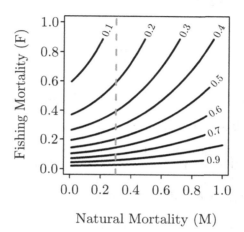

FIGURE 13.5. Contours of SPR values (values labeled at the edge of the contour line) for various values of natural and fishing mortality.

A more exact value of F that meets the SPR threshold given the value of M is found by searching for the last value greater than the threshold SPR value in the row from the SPRs matrix that corresponds to the value of M. The appropriate row is isolated by including the M value in quotes (because it is a label) within [].

```
> tmp <- SPRs["0.3",]
```

This search is aided by determining whether each value in the matrix is greater than the SPR threshold value. The F that corresponds to the last TRUE is the largest F that results in the threshold SPR being met.

```
> tmp>=0.4    # only first three rows shown
 0.01  0.02  0.03  0.04  0.05  0.06  0.07  0.08  0.09   0.1
 TRUE  TRUE  TRUE  TRUE  TRUE  TRUE  TRUE  TRUE  TRUE  TRUE
 0.11  0.12  0.13  0.14  0.15  0.16  0.17  0.18  0.19   0.2
 TRUE  TRUE  TRUE  TRUE  TRUE  TRUE  TRUE  TRUE  TRUE  TRUE
 0.21  0.22  0.23  0.24  0.25  0.26  0.27  0.28  0.29   0.3
 TRUE  TRUE  TRUE  TRUE  TRUE  TRUE  TRUE FALSE FALSE FALSE
```

Searching for the last TRUE in this vector, however, is still tedious work. This
search is aided by using which.min(), which returns the position of the first
FALSE in a vector. Given the nature of the SPR calculation, the last TRUE is in
the immediately previous position. The F value for this position is the column
name (found with colnames()) of the SPRs matrix in the same position.

```
> pos <- which.min(tmp>=0.4)
> colnames(SPRs)[pos-1]
[1] "0.27"
```

Thus, $F = 0.27$ or lower would meet the SPR threshold of 0.4 if $M = 0.3$ in
this fishery.

13.3   Year-Class Strength

Assessment of the strength of year-classes of fish is a key objective of many
fish monitoring activities. Year-class strength is often assessed by the catch
per unit effort (CPE) of juvenile prerecruit fishes in annual samplings. As
an example, Koenigs et al. (2015) used the catch of age-1 Walleye per trawl
tow (CPE) as an index of year-class strength of Walleye in Lake Winnebago
(Wisconsin; Figure 13.6). It is evident from Figure 13.6 that strong year-
classes were produced in 2001 and 2008 and that a series of poor year-classes
were produced in the late 1980s. This plot is constructed as described below.

The catch of yearling Walleye and number of trawl tows by year are in
WalleyeWyrlng.csv. The CPE is catch divided by number of tows. Koenigs
et al. (2015) rescaled the CPE values to have a mean of 0 and a standard deviation of 1. This particular rescaling is called *standardizing* and is accomplished
with scale() using the vector of values as the only argument.

```
> wae1yr <- read.csv("WalleyeWyrlng.csv") %>%
    mutate(cpe=yearlings/tows,scpe=scale(cpe))
> headtail(wae1yr)
   year yrclass yearlings tows        cpe        scpe
1  1986    1985        67  127 0.52755906  -0.5131486
2  1987    1986        55  138 0.39855072  -0.5531752
3  1988    1987         4  138 0.02898551  -0.6678380
23 2008    2007        60  137 0.43795620  -0.5409491
24 2009    2008      1257  129 9.74418605   2.3464394
25 2010    2009       109  138 0.78985507  -0.4317675
```

The plot of standardized CPE versus year-class with vertical lines from zero
for each CPE value (Figure 13.6) is constructed using type="h" in plot().
The width of the vertical lines is increased to look more like bars with lwd=
and the ends of the line are made square (rather than the default rounded)

with `lend=1`. The default x-axis is excluded with `xaxt="n"` and a manually constructed x-axis is added with two calls to `axis()`. First, slightly smaller than default tick marks (`tcl=` controls the tick length), but not labels, at each year-class are added. Second, "sideways" (i.e., `las=2`) labels are added at specific year-classes. Finally, a horizontal reference line is added at 0 with `abline()`.

```
> plot(scpe~yrclass,data=wae1yr,type="h",lwd=4,lend=1,
      xaxt="n",xlab="Year-Class",ylab="Standardized CPE")
> axis(1,at=wae1yr$yrclass,labels=NA,tcl=-0.1)
> axis(1,at=seq(1985,2005,5),las=2)
> abline(h=0)
```

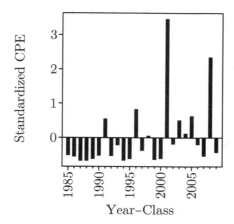

FIGURE 13.6. Plot of standardized catch of age-1 Lake Winnebago Walleye per trawl tow by year-class.

13.3.1   Catch-Curve Residuals

Indices of year-class strength derived from annual samplings of juvenile pre-recruit fish can provide critical information to fisheries managers. However, these types of monitoring activities are labor intensive, primarily because they must occur annually. The following sections provide alternative statistical methods for identifying strong year-classes from potentially less intensive sampling programs.

Maceina (1997) showed that the residuals from a catch curve (Section 11.2.3), potentially from one sample of catch-at-age data, could be used as an index of year-class strength. Year-classes with positive residuals are relatively stronger than year-classes with negative residuals. Maceina (1997) suggested using Studentized residuals, which are residuals standardized to have a standard deviation of 1. Because Studentized residuals follow a t distribution with

$n - 1$ degrees-of-freedom (here, n is the number of year-classes), the fisheries scientist can objectively define a "strong" or "weak" year-class based on where the residual is on the t distribution. For example, the fisheries scientists may call the year-class "strong" if the Studentized residual for that year-class is in the 80th percentile of the t distribution.

As an example of this method, Koenigs et al. (2015) captured adult Walleye from Lake Winnebago during spawning assessments in 2010. The sex was recorded and ages were estimated from sectioned otoliths for each fish. Koenigs et al. (2015) fit separate catch curves to female and male Walleye, but only female Walleye are used here.

```
> wae <- read.csv("WalleyeWad.csv")
> headtail(wae)
   age numF numM
1    2   NA   85
2    3    3  288
3    4   23  956
16  17   NA   26
17  18   NA   15
18  19    3   65
```

A weighted catch-curve was fit to age-5 (youngest fish fully recruited to the gear) through age-19 (oldest observed age) female Walleye with `catchCurve()` as described in Section 11.2.3.

```
> cc1 <- catchCurve(numF~age,data=wae,ages2use=5:19,
                    weighted=TRUE)
```

The actual ages used in the catch curve are contained in the `age.e` portion of the `catchCurve()` object. The year-class is computed by subtracting these ages from the capture year (2010). The Studentized residuals are extracted from the `lm` portion of the `catchCurve()` object with `rstudent()`. These data are combined into a data.frame below for convenience.

```
> res <- data.frame(age=cc1$age.e,yrclass=2010-cc1$age.e,
                    ycs=rstudent(cc1$lm))
> headtail(res,n=2)
   age yrclass        ycs
1    5    2005  0.2736657
2    6    2004 -0.3064128
14  18    1992         NA
15  19    1991  0.2885734
```

A plot of Studentized residuals versus year-class (Figure 13.7) is constructed similar to how Figure 13.6 was constructed, with the exception that the x-axis was manually constructed so that tick marks were not shown for year-classes not represented in the data.

```
> plot(ycs~yrclass,data=res,type="h",lwd=4,lend=1,xaxt="n",
       xlab="Year Class",ylab="Studentized Residual")
> axis(1,at=res$yrclass[!is.na(res$ycs)],labels=NA,tcl=-0.1)
> axis(1,at=c(1991,1996,2000,2004),las=2)
> abline(h=0)
```

This plot may be augmented with horizontal lines at critical Studentized residual values that identify "strong" and "weak" year-classes. For example, the Studentized residual that corresponds to the 20th and 80th percentiles of the t distribution is determined with qt(), which takes the desired *proportions* in a vector as the first argument and the degrees-of-freedom in df=.

```
> ( critres <- qt(c(0.20,0.80),df=length(res$ycs)-1) )
[1] -0.8680548  0.8680548
```

Horizontal lines at these critical Studentized residual values are added to the plot with abline().

```
> abline(h=critres,lty="dashed")
```

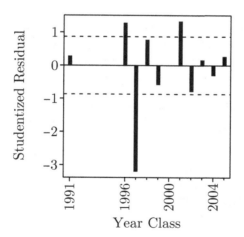

FIGURE 13.7. Studentized residuals from the weighted catch curve for age-5 to age-19 female Lake Winnebago Walleye. Horizontal dotted lines represent the upper and lower 20% of residuals. Year-classes above the upper dotted line are considered to be "strong" and year-classes below the lower dotted line are considered to be "weak."

From Figure 13.7 it appears that 1996 and 2001 were "strong" year-classes and 1997 was a "weak" year-class for female Walleye on Lake Winnebago.

13.3.2   Two-Way ANOVA Models

Kimura (1988) introduced and Maceina and Pereira (2007) demonstrated how differences in year-class strengths can be identified through the application of a two-way analysis of variance (ANOVA) model. A two-way ANOVA is an extension of the one-way ANOVA introduced in Section 8.3.2 to include two categorical explanatory variables in the model. Fitting the two-way ANOVA model allows one to determine if the mean of the response variable differs among levels of an explanatory variable *adjusted* for the levels of the other explanatory variable. In this application, the response variable is log CPE[17] and the two explanatory variables are age and year-class. Of particular interest from this analysis is the mean log CPE for each year-class *adjusted* for the effect of age on log CPE. Thus, the mean log CPE for each year class is adjusted for differential age-dependent catchabilities and can therefore be used as a relative measure of year-class strength (Maceina and Pereira 2007).

The data for this analysis is CPE recorded for each age from several years of samples. Year-class is added to the data by subtracting age from capture year. With these data there is one observation (CPE) per each age and year-class combination. This is important to note because most two-way ANOVA models include the interaction between the two explanatory variables. However, multiple observations are needed at each combination of the explanatory variables to fit an interaction. Thus, in this application, the two-way ANOVA model does not contain an interaction term.

As an example, Pratt et al. (2014) recorded the capture year and age of Lake Sturgeon captured in multiple gillnet sets in Goulais Bay, Lake Superior (Ontario) in July 2010–2012 (data in *SturgeonGB.csv*). The number of these fish in each age and year combination must be counted to generate the catch data required for the analysis. These counts are made with group_by() and summarize() from **dplyr** as described in Section 2.5.[18] A variable with year-class is added to the data.frame by subtracting the observed age from the capture year.

```
> d <- read.csv("SturgeonGB.csv") %>%
    group_by(year,age) %>%
    summarize(catch=n()) %>%
    as.data.frame() %>%
    mutate(yrclass=year-age)
> headtail(d)
   year age catch yrclass
1  2010   3    13    2007
2  2010   4    10    2006
3  2010   5    19    2005
25 2012   9     9    2003
26 2012  10    10    2002
27 2012  12     2    2000
```

The sample size per age and year-class combination is computed below with xtabs() (print() with zero.print="-" is used here to ease identification of the non-zero values in the table).

```
> print(xtabs(~yrclass+age,data=d),zero.print="-")
        age
yrclass 2 3 4 5 6 7 8 9 10 11 12
   2000 - - - - - - - -  1  1  1
   2001 - - - - - - - 1  1  -  -
   2002 - - - - - - 1 1  1  -  -
   2003 - - - - - 1 1 1  -  -  -
   2004 - - - - 1 1 1 -  -  -  -
   2005 - - - 1 1 1 - -  -  -  -
   2006 - - 1 1 1 - - -  -  -  -
   2007 - 1 1 1 - - - -  -  -  -
   2008 - 1 1 - - - - -  -  -  -
   2009 1 1 - - - - - -  -  -  -
```

These results illustrate that no more than one value exists at each age and year-class combination. However, two other common problems are also evident in these results. First, there are some age and year-class combinations where fish of that age should have been vulnerable to the gear but were not collected. For example, no age-11 fish were sampled from the 2001 year-class even though age-12 fish were sampled from the 2000 year-class. Additionally, no age-2 fish from the 2008 year-class were present, but age-2 fish from the 2009 year-class were present. In the current data.frame, these two age and year-class combinations are treated as missing values rather than as observations of zero fish, which will inflate the year-class strengths that would be computed from these values. This problem is addressed by adding catches of zero in these positions in the data.frame. Vectors with the appropriate capture year, age, a zero, and year-class are row-bound to the catch data.frame with rbind().

```
> d %<>% rbind(c(2012,11,0,2001),c(2010,2,0,2008))
> tail(d,n=3)
   year age catch yrclass
27 2012  12     2    2000
28 2012  11     0    2001
29 2010   2     0    2008
```

Finally, note that the 2009 year-class was represented in only the last two years of samples, whereas all other year-classes were represented in samples from all three capture years.

Pratt et al. (2014) used different numbers of nets in each year. Thus, to compute CPE, the number of nets fished per year (i.e., the effort) were entered into a small data.frame and then left_join() was used as described in Section 2.3.2 to appropriately (i.e., by year) append the effort values to the catch data.frame.

```
> eff <- data.frame(year=c(2010,2011,2012),effort=c(23,22,27))
> d %<>% left_join(eff,by="year")
> headtail(d)
   year age catch yrclass effort
1  2010   3    13    2007     23
2  2010   4    10    2006     23
3  2010   5    19    2005     23
27 2012  12     2    2000     27
28 2012  11     0    2001     27
29 2010   2     0    2008     23
```

As noted above, log CPE is used as the response variable in this analysis. However, before computing the CPE from `catch` and `effort`, it is important to note that log CPE will not exist for age and year-class combinations where the catch was zero. A common "fix" for this problem is to add 1 to the values before computing the logarithm (Gotelli and Ellison 2013). However, because the CPE values are so small in this example, adding a 1 to each CPE value is not inconsequential.[19] Thus, it is more appropriate to add 1 to the catches and then compute the CPE before taking the logarithms.

```
> d %<>% mutate(cpe=(catch+1)/effort,logcpe=log(cpe))
> headtail(d)
   year age catch yrclass effort        cpe     logcpe
1  2010   3    13    2007     23 0.60869565 -0.4964369
2  2010   4    10    2006     23 0.47826087 -0.7375989
3  2010   5    19    2005     23 0.86956522 -0.1397619
27 2012  12     2    2000     27 0.11111111 -2.1972246
28 2012  11     0    2001     27 0.03703704 -3.2958369
29 2010   2     0    2008     23 0.04347826 -3.1354942
```

Finally, both `age` and `yrclass` in this data.frame are quantitative, the explanatory variables in a two-way ANOVA must be categorical. Both variables are converted to factors below with `factor()`.

```
> d %<>% mutate(age=as.factor(age),yrclass=as.factor(yrclass))
```

The two-way ANOVA without an interaction term is fit using `lm()` with a model of the form `y~factor1+factor2` and the corresponding data.frame in `data=`. The ANOVA table is extracted from the `lm` object with `Anova()` from `car`.

```
> lm1 <- lm(logcpe~age+yrclass,data=d)
> Anova(lm1)
Anova Table (Type II tests)

Response: logcpe
          Sum Sq Df F value  Pr(>F)
age       15.6794 10 4.7615 0.01391
yrclass    6.5542  9 2.2115 0.12637
Residuals  2.9636  9
```

These results show that mean log CPE differs significantly among ages ($p = 0.0139$), which is not surprising, but not among year-classes ($p = 0.1264$).

Maceina and Pereira (2007) used *least-squares means* for each year-class as their metric of year-class strength. Least-squares means are computed in two steps (Lenth 2015). First, the two-way ANOVA model is used to predict the log CPE for each combination of age and year-class. The least-squares mean for one level is then the mean predicted log CPE computed across all levels of the other variable. For example, the least-squares mean for the 2000 year-class is the mean predicted log CPE for all ages and the 2000 year-class. Similarly, the least-squares mean for age-3 is the mean predicted log CPE for age-3 and all year-classes.[20]

Least-squares means are computed from the lm() object with lsmeans() from **lsmeans** (Lenth and Herve 2015). This function requires the explanatory variable for which to compute the least-squares means in a formula as the second argument. The least-squares means and confidence intervals adjusted for multiple comparisons using the Tukey HSD method (see Section 8.3.2) are extracted from the lsmeans() object with summary().

```
> yc.lsm <- lsmeans(lm1,~yrclass)
> ( yce <- summary(yc.lsm) )
 yrclass      lsmean         SE df    lower.CL     upper.CL
   2000   1.09081120 0.6713557  9 -0.4279009   2.60952332
   2001  -0.27693502 0.5859899  9 -1.6025363   1.04866626
   2002  -0.03557849 0.5149889  9 -1.2005643   1.12940735
   2003  -1.04356730 0.4700166  9 -2.1068186   0.01968402
   2004  -1.65481821 0.4536063  9 -2.6809470  -0.62868941
   2005  -1.85245846 0.4698917  9 -2.9154274  -0.78948950
   2006  -2.32550488 0.5153683  9 -3.4913490  -1.15966077
   2007  -1.82755374 0.5838159  9 -3.1482370  -0.50687045
   2008  -1.54570466 0.6672506  9 -3.0551304  -0.03627891
   2009  -0.47217266 0.7722611  9 -2.2191486   1.27480333
```

Results are averaged over the levels of: age
Confidence level used: 0.95

Tukey-adjusted tests for differences between each pair of means and letters indicating groups that are statistically similar are constructed by submitting the lsmeans() object to pairs() and cld() from **lsmeans**, respectively. Because the year-class effect was not significant in this example, the results from these functions are not shown here.

```
> pairs(yc.lsm)
> cld(yc.lsm)
```

The least-squares means are plotted against year-class to produce a visual of year-class strength (Figure 13.8). This plot is constructed with plotCI() from **plotrix** as described in Section 3.5.2. Note, however, in the code below that with() is used to reduce the repetitive typing of yce$. Further note that the yrclass variable in yce must be converted to numeric values with fact2num() from **FSA** for plotting purposes.

```
> with(yce,plotCI(fact2num(yrclass),lsmean,
                li=lower.CL,ui=upper.CL,pch=19,cex=0.7,
                xlab="Year-Class",ylab="Strength Index"))
> with(yce,lines(fact2num(yrclass),lsmean))
```

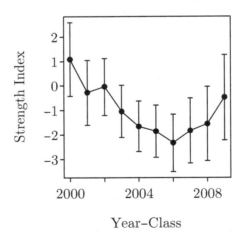

FIGURE 13.8. Least-squares means, with Tukey-adjusted 95% confidence intervals, of log CPE of Goulais Bay Lake Sturgeon by year-class.

Finally, the assumptions for the two-way ANOVA, which are the same as for the one-way ANOVA, should be assessed with a modified residual plot and a histogram of residuals (Figure 13.9; see Section 8.3.2.1). The default for the modified residual plot from residPlot() is to show boxplots of residuals for each group. However, this is not appropriate here because each group consists of only one observation. Thus, `bp=FALSE` is used to plot points rather than boxplots for each group. The residual plot shows a curvature (largely dictated by three points with low fitted values) but no obvious heteroscedasticity. The histogram of residuals is primarily symmetric. The assumptions for the two-way ANOVA were better met when fish from the 2009 year-class, which were represented in only two sample years, were removed from the analysis (results not shown).

```
> residPlot(lm1,bp=FALSE)
```

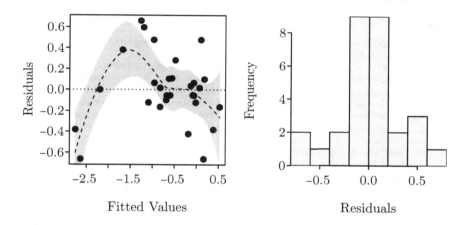

FIGURE 13.9. Modified residual plot (Left) and histogram of residuals (Right) from fitting a two-way ANOVA without an interaction term to the log CPE of Goulais Bay Lake Sturgeon by age and year-class.

13.4 Further Considerations

Much research has examined the stock-recruit relationship and the models employed to describe those relationships. The methods discussed in Section 13.1 only touch on the very beginning of this research. The interested reader should consult Subbey et al. (2014) for a careful summary of current knowledge regarding modeling stock-recruit relationships. Their summary included important discussions about different types of stock-recruit models, incorporating environmental effects, different error structures (i.e., not just normal or lognormal), bias and bias correction when fitting with lognormal errors, the use of AIC/BIC in model selection, and the use of other model fitting paradigms (nonparametric smoothers (e.g., Cadigan (2013), who included example R code) and hierarchical Bayesian models).

Spawning stock biomass and eggs per recruit analyses are used to evaluate the effects of fishing mortality and age-at-first-capture on the spawning potential of a stock (Gabriel et al. 1989). Thus, these analyses have a broad goal that is similar to the SPR analysis of Section 13.2. These methods are implemented in sbpr() and epr() from **fishmethods**.

Notes

[1]Data manipulations in this chapter require functions from **magrittr** and **dplyr**, which are fully described in Chapter 2.

[2]Derivations of the Beverton-Holt function are found in Mangel (2006) and Haddon (2011).

[3]Models can often be rewritten to have different parameters. These different versions are called *parameterizations*. Different parameterizations of the same model fit the data exactly the same way (i.e., predicted values of the response (or dependent) variable are equivalent) and have the same number of parameters (i.e., they are not more parsimonious). Different parameterizations are used to estimate different parameters, though there may also be algorithmic considerations.

[4]Derivations of the Ricker function are found in Mangel (2006) and Haddon (2011).

[5]The linearized form of the Beverton-Holt function is found by inverting both sides of the equation and then simplifying the right-hand side.

[6]The linearized form of the Ricker function is found by dividing both sides by S and then taking the logarithm of both sides.

[7]The srStarts() function can be used to derive starting values for other parameterizations of the Beverton-Holt and Ricker stock-recruitment functions. These other parameterizations may be seen with srModels() (no arguments). Starting values for these other parameterizations may be obtained by submitting the appropriate name for the parameterization in param= in srStarts().

[8]This use of dynamic plots requires the **relax, tcltk,** and **tkrplot** packages which may require manual installation in some environments. See Section 1.3 for instructions on how to manually install packages in R.

[9]Quinn II and Deriso (1999, p. 104) stated that the theory used to develop the Beverton-Holt and Ricker functions suggested using multiplicative errors and that the multiplicative error model "fits the error structure of actual data sets fairly well."

[10]The assumptions of nonlinear regression are assessed as described in Section 12.3.6.

[11]Other functions can be entered manually with function() as described in Section 1.8.2.

[12]Similar to the previously used Ricker function, a density-independent function is created with srFuns() and starting values are obtained with srStarts(). The starting value for a (the only parameter in the density-independent function) is obtained in srStarts() by fitting a simple linear regression of R on S through the origin (i.e., without an intercept).

[13]Note the use of [[]] in this code so that the results will not retain the original, and ultimately inappropriate, label. See Section 1.7.1.

[14]Many authors will treat the fecundity variable as the mean fecundity for all females, rather than just the mature females as described here. If that is the case, then the E_i of those authors is the same as $\mu_i E_i$ used here.

[15]In this scenario, it would be more efficient to put the constant values of M and F into rep() rather than storing as separate objects. However, the simulations in the next section are more efficient with M and F as separate objects. Thus, the use of separate objects is illustrated here.

[16]The vulnerability of fish to the fishery is usually length- rather than age-dependent. Due

to variability in length within age-classes, most fisheries will have some ages where some fish are and some fish are not vulnerable to the fishery.

[17]Kimura (1988) argued that catch data should be log transformed to stabilize the variances (i.e., meet the homoscedasticity assumption). Hubert and Fabrizio (2007) noted that many distributions of CPE are right-skewed and further noted that many fisheries scientists will log-transform CPE data to achieve normality. However, Hubert and Fabrizio (2007, p. 297) stated "[i]t is our experience that logarithmic transformations of [CPE] data seldom yield a normal distribution but can reduce the variance relative to the mean."

[18]Note that the grouping structure must be removed from the data.frame with `as.data.frame()` so that the data.frame can be further modified.

[19]A "small" value could be added to the CPE data. As a general rule-of-thumb, the value added should be no larger than the smallest value that could be observed for the variable. In this case, a value no larger than 1 (the smallest possible non-zero catch) divided by the maximum effort value could be added to CPE.

[20]The predictions at each age and year-class combination are seen with `summary(ref.grid(lm1))` from **lsmeans**.

References

Aho, K., Derryberry, D., and Peterson, T. (2014). Model selection for ecologists: The worldviews of AIC and BIC. *Ecology*, 95:631–636.

Aho, K. A. (2014). *Foundational and Applied Statistics for Biologists Using R*. Chapman & Hall/CRC, Boca Raton, FL.

Aho, K. A. (2015a). *asbio: A Collection of Statistical Tools for Biologists*. R package.

Aho, K. A. (2015b). *Foundational and Applied Statistics for Biologists Using R: Electronic Appendix (Introduction to R)*. Chapman & Hall/CRC, Boca Raton, FL. Accessed from http://www2.cose.isu.edu/~ahoken/book/Intro_to_R.pdf on June 18, 2015.

Allen, M. S. and Hightower, J. E. (2010). Fish population dynamics: Mortality, growth, and recruitment. In Hubert, W. A. and Quist, M. C., editors, *Inland Fisheries Management in North America*, chapter 2, pages 43–79. American Fisheries Society, Bethesda, MD, third edition.

Altman, D. G. and Bland, J. M. (1983). Measurements in medicine: The analysis of method comparison studies. *The Statistician*, 32:307–317.

Anderson, R. O. (1976). Management of small warm water impoundments. *Fisheries*, 1(6):5–7, 26–28.

Bache, S. M. and Wickham, H. (2014). *magrittr: A Forward-Pipe Operator for R*. R package.

Bailey, N. T. J. (1951). On estimating the size of mobile populations from capture-recapture data. *Biometrika*, 38:293–306.

Bailey, N. T. J. (1952). Improvements in the interpretation of recapture data. *Journal of Animal Ecology*, 21:120–127.

Baillargeon, S. and Rivest, L.-P. (2007). Rcapture: Loglinear models for capture-recapture in R. *Journal of Statistical Software*, 19(5):1–31.

Barada, T. J. (2009). Catfish population dynamics in the Platte River, Nebraska. Master's thesis, University of Nebraska, Lincoln, NE.

Bates, D. M., Maechler, M., Bolker, B. M., and Walker, S. (2014). *lme4: Linear Mixed-Effects Models using Eigen and S4*. R package.

Bates, D. M. and Watts, D. G. (2007). *Nonlinear Regression Analysis and Its Applications*. John Wiley & Sons, Inc., Hoboken, NJ.

Baty, F., Ritz, C., Charles, S., Brutsche, M., Flandrois, J.-P., and Delignette-Muller, M.-L. (2015). A toolbox for nonlinear regression in R: The package nlstools. *Journal of Statistical Software*, 66(5):1–21.

Beamish, R. J. and Fournier, D. A. (1981). A method for comparing the precision of a set of age determinations. *Canadian Journal of Fisheries and Aquatic Sciences*, 38:982–983.

Berg, C. W. and Kristensen, K. (2012). Spatial age-length key modelling using continuation ratio logits. *Fisheries Research*, 129–130:119–126.

Bettoli, P. W. and Miranda, L. E. (2001). A cautionary note about estimating mean length at age with subsampled data. *North American Journal of Fisheries Management*, 21:425–428.

Beverton, R. J. H. and Holt, S. J. (1957). *On the Dynamics of Exploited Fish Populations*, volume 19 of *Fisheries Investigations (Series 2)*. United Kingdom Ministry of Agriculture and Fisheries.

Birchell, G. J. (2007). The effects of invasive Brook Trout removal on native Colorado River Cutthroat Trout in a small headwater stream in Northeastern Utah. Master's thesis, Utah State University.

Blackwell, B. G., Brown, M. L., and Willis, D. W. (2000). Relative weight (Wr) status and current use in fisheries assessment and management. *Reviews in Fisheries Science*, 8:1–44.

Bolger, T. and Connolly, P. L. (1989). The selection of suitable indices for the measurement and analysis of fish condition. *Journal of Fish Biology*, 34:171–182.

Bolker, B. M. (2008). *Ecological Models and Data in R*. Princeton University Press, Princeton, NJ.

Bolker, B. M., Gardner, B., Maunder, M., Berg, C. W., Brooks, M., Comita, L., Crone, E., Cubaynes, S., Davies, T., deValpine, P., Ford, J., Gimenez, O., Kery, M., Kim, E. J., Lennart-Cody, C., Magnusson, A., Martell, S., Nash, J., Nielsen, A., Regetz, J., Skaug, H., and Zipkin, E. (2013). Strategies for fitting nonlinear ecological models in R, AD Model Builder, and BUGS. *Methods in Ecology and Evolution*, 4:501–512.

Bolker, B. M. and R Development Core Team (2014). *bbmle: Tools for General Maximum Likelihood Estimation*. R package.

Borchers, D. L., Stephens, S., and Zucchini, W. (2004). *Estimating Animal Abundance: Closed Populations*. Springer, New York, NY.

Borchers, H. W. (2015). *pracma: Practical Numerical Math Functions*. R package.

Bowker, A. H. (1948). A test for symmetry in contingency tables. *Journal of the American Statistical Association*, 43:572–574.

Brenden, T. O., Murphy, B. R., and Birch, J. B. (2003). Statistical properties of the relative weight (Wr) index and an alternative procedure for testing Wr differences between groups. *North American Journal of Fisheries Management*, 23:1136–1151.

Brenden, T. O., Wagner, T., and Murphy, B. R. (2008). Novel tools for analyzing proportional size distribution index data. *North American Journal of Fisheries Management*, 28:1233–1242.

Burnham, K. P. and Anderson, D. R. (2002). *Model Selection and Multimodel Inference: A Practical Information-Theoretic Approach*. Springer, New York, NY, second edition.

Burnham, K. P. and Anderson, D. R. (2004). Multimodel inference: Understanding AIC and BIC in model selection. *Sociological Methods and Research*, 33:261–304.

Cade, B. S. and Noon, B. R. (2003). A gentle introduction to quantile regression for ecologists. *Frontiers in Ecology and the Environment*, 1:412–420.

Cade, B. S., Terrell, J. W., and Neely, B. C. (2011). Estimating geographic variation on allometric growth and body condition of Blue Suckers with quantile regression. *Transactions of the American Fisheries Society*, 140:1657–1669.

Cadigan, N. G. (2013). Fitting a non-parametric stock-recruitment model in R that is useful for deriving MSY reference points and accounting for model uncertainty. *ICSE Journal of Marine Science*, 70:56–67.

Cailliet, G. M., Smith, W. D., Mollet, H. F., and Goldman, K. J. (2006). Age and growth studies of chondrichthyan fishes: The need for consistency in terminology, verification, validation, and growth function fitting. *Environmental Biology of Fish*, 77:211–228.

Campana, S. E. (2001). Accuracy, precision and quality control in age determination, including a review of the use and abuse of age validation methods. *Journal of Fish Biology*, 59:197–242.

Campana, S. E., Annand, M. C., and McMillan, J. I. (1995). Graphical and statistical methods for determining the consistency of age determinations. *Transactions of the American Fisheries Society*, 124:131–138.

Carle, F. L. and Strub, M. R. (1978). A new method for estimating population size from removal data. *Biometrics*, 34:621–630.

Carstensen, B., Gurrin, L., Ekstrom, C., and Figurski, M. (2015). *MethComp: Functions for Analysis of Agreement in Method Comparison Studies*. R package.

Chan, C., Chan, G. C., and Leeper, T. J. (2015). *rio: A Swiss-army knife for data file I/O*. R package.

Chang, W. (2012). *R Graphics Cookbook: Practical Recipes for Visualizing Data*. O'Reilly Media, Sebastopol, CA.

Chang, W. Y. B. (1982). A statistical method for evaluating the reproducibility of age determination. *Canadian Journal of Fisheries and Aquatic Sciences*, 39:1208–1210.

Chao, A. and Huggins, R. M. (2005). Modern closed-population capture-recapture models. In Amstrup, S. C., McDonald, T. L., and Manly, B. F. J., editors, *Handbook of Capture-Recapture Analysis*, chapter 4, pages 58–87. Princeton University Press, Princeton, NJ.

Chapman, D. G. (1951). Some properties of the hypergeometric distribution with applications to zoological censuses. *University of California Publications on Statistics*, 1:131–160.

Chapman, D. G. and Robson, D. S. (1960). The analysis of a catch curve. *Biometrics*, 16:354–368.

Cloern, J. E. and Nichols, F. H. (1978). A von Bertalanffy growth model with a seasonally varying coefficient. *Journal of the Fisheries Research Board of Canada*, 35:1479–1482.

Coggins Jr., L. G., Gwinn, D. C., and Allen, J. S. (2013). Evaluation of age-length key sample sizes required to estimate fish total mortality and growth. *Transactions of the American Fisheries Society*, 142:832–840.

Cone, R. S. (1989). The need to reconsider the use of condition indices in fishery science. *Transactions of the American Fisheries Society*, 118:510–514.

Conway, J., Eddelbuettel, D., Nishiyama, T., Prayaga, S. K., and Tiffin, N. (2013). *RPostgreSQL: R Interface to the PostgreSQL Database System*. R package.

Cooney, P. B. and Kwak, T. J. (2010). Development of standard weight equations for Caribbean and Gulf of Mexico amphidromous fishes. *North American Journal of Fisheries Management*, 30:1203–1209.

Cope, J. M. and Punt, A. E. (2007). Admitting ageing error when fitting growth curves: An example using the von Bertalanffy growth function with random effects. *Canadian Journal of Fisheries and Aquatic Sciences*, 64:205–218.

Cormack, R. M. (1989). Loglinear models for capture-recapture. *Biometrics*, 45:395–413.

Dalgaard, P. (2002). *Introductory Statistics with R*. Springer, New York, NY.

DeLury, D. B. (1947). On the estimation of biological populations. *Biometrics*, 3:145–167.

deVries, A. and Meys, J. (2012). *R for Dummies*. John Wiley & Sons, Inc., Hoboken, NJ.

Dinno, A. (2015). *dunn.test: Dunn's Test of Multiple Comparisons Using Rank Sums*. R package.

Dowle, M., Short, T., Lianoglou, S., and Srinivasan, A. (2014). *data.table: Extension of data.frame*. R package.

Dragulescu, A. A. (2014). *xlsx: Read, Write, Format Excel 2007 and Excel 97/2000/XP/2003 Files*. R package.

Dunn, A., Francis, R. I. C. C., and Doonan, I. J. (2002). Comparison of the Chapman-Robson and regression estimators of Z from catch-curve data when non-sampling stochastic error is present. *Fisheries Research*, 59:149–159.

Dunn, O. J. (1964). Multiple comparisons using rank sums. *Technometrics*, 6:241–252.

Efford, M. (2015). *secr: Spatially Explicit Capture-Recapture Models*. R package.

Elzhov, T. V., Mullen, K. M., Spiess, A.-N., and Bolker, B. M. (2013). *minpack.lm: R Interface to the Levenberg-Marquardt Nonlinear Least-Squares Algorithm Found in MINPACK, Plus Support for Bounds*. R package.

Escati-Penaloza, G., Parma, A. M., and Orensanz, J. M. (2010). Analysis of longitudinal growth increment data using mixed-effects models: Individual and spatial variability in a clam. *Fisheries Research*, 105:91–101.

Evans, G. T. and Hoenig, J. M. (1998). Testing and viewing symmetry in contingency tables, with application to readers of fish ages. *Biometrics*, 54:620–629.

Fabens, A. J. (1965). Properties and fitting of the von Bertalanffy growth curve. *Growth*, 29:265–289.

Faraway, J. J. (2005). *Extending the Linear Model with R: Generalized Linear, Mixed Effects and Nonparametric Regression Models*. Chapman & Hall/CRC, Boca Raton, FL.

Fienberg, S. E. (1980). *The Analysis of Cross-Classified Categorical Data.* MIT Press, Cambridge, MA, second edition.

Fiske, I. J. and Chandler, R. B. (2011). unmarked: An R package for fitting hierarchical models of wildlife occurrence and abundance. *Journal of Statistical Software*, 43(10):1–23.

Ford, E. (1933). An account of the Herring investigations conducted at Plymouth during the years from 1924–1933. *Journal of the Marine Biology Association (U.K.)*, 19:305–384.

Fox, J. (1997). *Applied Regression Analysis, Linear Models, and Related Methods.* Sage Publications, Thousand Oaks, CA.

Fox, J. (2008). *Applied Regression Analysis and General Linear Models.* Sage Publications, Thousand Oaks, CA, second edition.

Fox, J. and Weisberg, S. (2010). Nonlinear regression and nonlinear least squares in R. An Appendix to *An R Companion to Applied Regression*, second edition. Accessed from http://socserv.mcmaster.ca/jfox/Books/ Companion/appendix/Appendix-Nonlinear-Regression.pdf on June 18, 2015.

Fox, J. and Weisberg, S. (2011). *An R Companion to Applied Regression.* Sage Publications, Thousand Oaks, CA, second edition.

Francis, R. I. C. C. (1988). Are growth parameters estimated from tagging and age-length data comparable? *Canadian Journal of Fisheries and Aquatic Sciences*, 45:936–942.

Froese, R. (2006). Cube law, condition factor and weight-length relationships: History, meta-analysis and recommendations. *Journal of Applied Ichthyology*, 22:241–253.

Gabelhouse, D. W. (1984). A length-categorization system to assess fish stocks. *North American Journal of Fisheries Management*, 4:273–285.

Gabriel, W. L., Sissenwine, M. P., and Overholtz, W. J. (1989). Analysis of spawning stock biomass per recruit: An example for Georges Bank Haddock. *North American Journal of Fisheries Management*, 9:383–391.

Gandrud, C. (2014). *Reproducible Research with R and RStudio.* Chapman & Hall/CRC, Boca Raton, FL.

Garrett, R. G. (2015). *rgr: Applied Geochemistry EDA.* R package.

Gascuel, D. (1994). Une methode simple d'ajustement des cles taille/age: Application aux captures d'albacores (*Thunnus albacores*) de l'Atlantique est. *Canadian Journal of Fisheries and Aquatic Sciences*, 51:723–733.

Gerow, K. G. (2010). Biases with the regression line percentile method and the fallacy of a single standard weight. *North American Journal of Fisheries Management*, 30:679–690.

Gerow, K. G. (2011). Comment: Assessing length-related biases in standard weight equations. *North American Journal of Fisheries Management*, 31:656–660.

Gerow, K. G., Anderson-Sprecher, R. C., and Hubert, W. A. (2005). A new method to compute standard-weight equations that reduces length-related bias. *North American Journal of Fisheries Management*, 25:1288–1300.

Gerow, K. G., Hubert, W. A., and Anderson-Sprecher, R. C. (2004). An alternative approach to detection of length-related biases in standard weight equations. *North American Journal of Fisheries Management*, 24:903–910.

Gerritsen, H. D., McGrath, D., and Lordan, C. (2006). A simple method for comparing age length keys reveals significant regional differences within a single stock of Haddock (*Melanogrammus aeglefinus*). *ICES Journal of Marine Science*, 63:1096–1100.

Gompertz, B. (1825). On the nature of the function expressive of the law of human mortality and on a new mode of determining the value of life contingencies. *Philosophical Transactions of the Royal Society of London*, 115:515–585.

Goodyear, C. P. (1989). Spawning stock biomass per recruit: The biological basis for a fisheries management tool. Technical Report SCRS/89/82, ICCAT Working Document.

Goodyear, C. P. (1993). Spawning stock biomass per recruit in fisheries management: Foundation and current use. *Canadian Special Publication in Fisheries and Aquatic Sciences*, 120:67–81.

Goodyear, C. P. (1995). Mean size at age: An evaluation of sampling strategies with simulated Red Grouper data. *Transactions of the American Fisheries Society*, 124:746–755.

Gotelli, N. J. and Ellison, A. M. (2013). *A Primer of Ecological Statistics*. Sinauer Associates, Sunderland, MA, second edition.

Grothendieck, G. (2013). *nls2: Non-Linear Regression with Brute Force*. R package.

Gustafson, K. A. (1988). Approximating confidence intervals for indices of fish population size structure. *North American Journal of Fisheries Management*, 8:139–141.

Gutreuter, S. and Krzoska, D. J. (1994). Quantifying precision of *in situ* length and weight measurements of fish. *North American Journal of Fisheries Management*, 14:318–322.

Guy, C. S. and Brown, M. L. (2007). *Analysis and Interpretation of Freshwater Fisheries Data*. American Fisheries Society, Bethesda, MD.

Guy, C. S., Neumann, R. M., and Willis, D. W. (2006). New terminology for proportional stock density (PSD) and relative stock density (RSD): Proportional size structure (PSS). *Fisheries*, 31:86–87.

Guy, C. S., Neumann, R. M., Willis, D. W., and Anderson, R. O. (2007). Proportional size distribution (PSD): A further refinement of population size structure index terminology. *Fisheries*, 32:348.

Haddon, M., Mundy, C., and Tarbath, D. (2008). Using an inverse-logistic model to describe growth increments of Blacklip Abalone (*Haliotis rubra*) in Tasmania. *Fishery Bulletin*, 106:58–71.

Haddon, M. J. (2011). *Modelling and Quantitative Methods in Fisheries*. Chapman & Hall/CRC, Boca Raton, FL, second edition.

Harding, R. D., Hoover, C. L., and Marshall, R. P. (2010). Abundance of Cutthroat Trout in Auke Lake, Southeast Alaska, in 2005 and 2006. Technical report, Alaska Department of Fish and Game Fisheries Data Series No. 10-82.

Harrell, F. E. (2015). *Hmisc: Harrell Miscellaneous*. R package.

Hayes, D. B., Bence, J. R., Kwak, T. J., and Thompson, B. E. (2007). Abundance, biomass, and production. In Guy, C. S. and Brown, M. L., editors, *Analysis and Interpretation of Freshwater Fisheries Data*, chapter 8, pages 327–374. American Fisheries Society, Bethesda, MD.

He, J. X. and Bence, J. R. (2007). Modeling annual growth variation using a hierarchical Bayesian approach and the von Bertalanffy growth function, with application to Lake Trout in southern Lake Huron. *Transactions of the American Fisheries Society*, 136:318–330.

He, J. X., Bence, J. R., Johnson, J. E., Clapp, D. F., and Ebener, M. P. (2008). Modeling variation in mass-length relations and condition indices of Lake Trout and Chinook Salmon in Lake Huron: A hierarchial Bayesian approach. *Transactions of the American Fisheries Society*, 137:801–817.

Hedger, R. D., de Eyto, E., Dillane, M., Diserud, O. H., Hindar, K., McGinnity, P., Poole, R., and Rogan, G. (2013). Improving abundance estimates from electrofishing removal sampling. *Fisheries Research*, 137:104–115.

Helidoniotis, F., Haddon, M., Tuck, G., and Tarbath, D. (2011). The relative suitability of the von Bertalanffy, Gompertz and inverse logistic models for describing growth in Blacklip Abalone populations (*Haliotis rubra*) in Tasmania Australia. *Fisheries Research*, 112:12–31.

Helser, T. E. and Lai, H.-L. (2004). A Bayesian hierarchical meta-analysis of fish growth: With an example for North American Largemouth Bass, *Micropterus salmoides*. *Ecological Modelling*, 178:399–416.

Henderson, B. A. (1985). Factors affecting growth and recruitment of Yellow Perch, *Perca flavescens* Mitchill, in South Bay, Lake Huron. *Journal of Fisheries Biology*, 26:449–458.

Hightower, J. E. and Pollock, K. H. (2013). Tagging methods for estimating population size and mortality rates of inland Striped Bass populations. *American Fisheries Society Symposium*, 80:249–262.

Hilborn, R. and Mangel, M. (1997). *The Ecological Detective: Confronting Models with Data*. Princeton University Press, Princeton, NJ.

Hilborn, R. and Walters, C. J. (2001). *Quantitative Fisheries Stock Assessment: Choice, Dynamics, & Uncertainty*. Chapman & Hall, New York, NY, second edition.

Hoenig, J. M. (1983). Empirical use of longeity data to estimate mortality rates. *U.S. National Marine Fisheries Service Fishery Bulletin*, 81:898–903.

Hoenig, J. M., Barrowman, N. J., Hearn, W. S., and Pollock, K. H. (1998). Multiyear tagging studies incorporating fishing effort data. *Canadian Journal of Fisheries and Aquatic Sciences*, 55:1466–1476.

Hoenig, J. M. and Heisey, D. M. (1987). Use of a log-linear model with the EM algorithm to correct estimates of stock composition and to convert length to age. *Transactions of the American Fisheries Society*, 116:232–243.

Hoenig, J. M., Heisey, D. M., and Hanumara, R. C. (1993). Using prior and current information to estimate age composition: A new kind of age-length key. C.M. Document D:52, International Council for Exploration of the Sea.

Hoenig, J. M., Heisey, D. M., and Hanumara, R. C. (1994). A computationally simple approach to using current and past data in an age-length key. C.M. Document D:10, International Council for Exploration of the Sea.

Hoenig, J. M., Lawing, W. D., and Hoenig, N. A. (1983). Using mean age, mean length and median length data to estimate the total mortality rate. C.M. Document D:23, International Council for Exploration of the Sea, Copenhagen, Denmark.

Hoenig, J. M., Morgan, M. J., and Brown, C. A. (1995). Analysing differences between two age determination methods by tests of symmetry. *Canadian Journal of Fisheries and Aquatic Sciences*, 52:364–368.

Hollander, M., Wolfe, D. A., and Chicken, E. (2014). *Nonparametric Statistical Methods*. John Wiley & Sons, Inc., Hoboken, NJ, third edition.

Holm, S. (1979). A simple sequentially rejective multiple test procedure. *Scandinavian Journal of Statistics*, 6:65–70.

Hothorn, T., Bretz, F., and Westfall, P. (2008). Simultaneous inference in general parametric models. *Biometrical Journal*, 50:346–363.

Hubert, W. A. and Fabrizio, M. C. (2007). Relative abundance and catch per unit effort. In Guy, C. S. and Brown, M. L., editors, *Analysis and Interpretation of Freshwater Fisheries Data*, chapter 7, pages 279–325. American Fisheries Society, Bethesda, MD.

Isermann, D. A. and Knight, C. T. (2005). A computer program for age-length keys incorporating age assignment to individual fish. *North American Journal of Fisheries Management*, 25:1153–1160.

Isermann, D. A. and Paukert, C. P. (2010). Regulating harvest. In Hubert, W. A. and Quist, M. C., editors, *Inland Fisheries Management in North America*, chapter 7, pages 185–212. American Fisheries Society, Bethesda, MD, third edition.

Jakob, E. M., Marshal, S. D., and Uetz, G. W. (1996). Estimating fitness: A comparison of body condition indices. *Oikos*, 77:61–67.

Jellyman, P. G., Booker, D. J., Crow, S. K., and Jellyman, D. J. (2013). Does one size fit all? An evaluation of length-weight relationships for New Zealand's freshwater fish species. *New Zealand Journal of Marine and Freshwater Research*, 47:450–468.

Jennings, C. A., Sloss, B. L., Lasee, B. A., Burtle, G. J., and Moyer, G. R. (2012). Care, handling, and examination of sampled organisms. In Zale, A. V., Parrish, D. L., and Sutton, T. M., editors, *Fisheries Techniques, Third Edition*, chapter 5, pages 163–221. American Fisheries Society, Bethesda, MD.

Jiang, H., Pollock, K. H., Brownie, C., Hoenig, J. M., Latour, R. J., Wells, B. K., and Hightower, J. E. (2007). Tag return models allowing for harvest and catch and release: Evidence of environmental and management impacts on Striped Sass fishing and natural mortality rates. *North Amercian Journal of Fisheries Management*, 27:387–396.

Jolly, G. M. (1965). Explicit estimates from capture-recapture data with both death and immigration – Stochastic model. *Biometrika*, 52:225–247.

Jones, C. M. (2000). Fitting growth curves to retrospective size-at-age data. *Fisheries Research*, 46:123–129.

Katsanevakis, S. and Maravelias, C. D. (2008). Modelling fish growth: Multi-model inference as a better alternative to *a priori* using von Bertalanffy equation. *Fish and Fisheries*, 9:178–187.

Kenchington, T. J. (2014). Natural mortality estimators for information-limited fisheries. *Fish and Fisheries*, 15:533–562.

Kimura, D. K. (1977). Statistical assessment of the age-length key. *Journal of the Fisheries Research Board of Canada*, 34:317–324.

Kimura, D. K. (1980). Likelihood methods for the von Bertalanffy growth curve. *Fishery Bulletin*, 77:765–776.

Kimura, D. K. (1988). Analyzing relative abundance indices with log-linear models. *Transactions of the American Fisheries Society*, 8:175–180.

Kimura, D. K. (2000). Using nonlinear functional relationship regression to fit fisheries models. *Canadian Journal of Fisheries and Aquatic Sciences*, 57:160–170.

Kimura, D. K. (2008). Extending the von Bertalanffy growth model using explanatory variables. *Canadian Journal of Fisheries and Aquatic Sciences*, 65:1879–1891.

Kimura, D. K. and Anderl, D. M. (2005). Quality control of age data at the Alaska Fisheries Science Center. *Marine and Freshwater Research*, 56:783–789.

Kimura, D. K. and Chikuni, S. (1987). Mixtures of empirical distributions: An iterative application of the age-length key. *Biometrics*, 43:23–35.

Knight, W. (1968). Asymptotic growth: An example of nonsense disguised as mathematics. *Journal of the Fisheries Research Board of Canada*, 25:1303–1307.

Koch, J. D., Neely, B. C., and Colvin, M. E. (2014). Evaluation of precision and sample sizes using standardized sampling in Kansas reservoirs. *North American Journal of Fisheries Management*, 34:1211–1220.

Koenigs, R. P., Bruch, R. M., Stelzer, R. S., and Kamke, K. K. (2015). Validation of otolith ages for Walleye (*Sander vitreus*) in the Winnebago System. *Fisheries Research*, 167:13–21.

Koenker, R. (2015). *quantreg: Quantile Regression*. R package.

Krebs, C. J. (1999). *Ecological Methodology*. Addison-Welsey Educational Publishing, Menlo Park, CA, second edition.

Kristensen, K. and Berg, C. (2010). *DATRAS: Read and Convert Raw Data Obtained from the DATRAS Database of Trawl Surveys.* R package.

Laake, J., Borchers, D., Thomas, L., Miller, D., and Bishop, J. (2015). *mrds: Mark-Recapture Distance Sampling.* R package.

Laake, J. L. (2013). RMark: An R interface for analysis of capture-recapture data with MARK. AFSC Processed Report 2013-01, Alaska Fisheries Science Center, NOAA, National Marine Fisheries Service, Seattle, WA.

Laake, J. L., Johnson, D. S., and Conn, P. B. (2013). marked: an R package for maximum likelihood and Markov Chain Monte Carlo analysis of capture-recapture data. *Methods in Ecology and Evolution*, 4:885–890.

Lai, H.-L. (1987). Optimum allocation for estimating age composition using age-length key. *Fishery Bulletin*, 85:179–185.

Lai, H.-L. (1993). Optimum sampling design for using the age-length key to estimate age composition of a fish population. *Fishery Bulletin*, 92:382–388.

Lai, H.-L. and Helser, T. (2004). Linear mixed-effects models for weight-length relationships. *Fisheries Research*, 70:377–387.

Lamport, L. (1994). *LaTeX: A document preparation system: User's guide and reference.* Addison-Wesley, Menlo Park, CA.

Le Cren, E. D. (1951). The length-weight relationship and seasonal cycle in gonad weight and condition in the Perch (*Perca flavescens*). *Journal of Animal Ecology*, 20:201–219.

Lehnert, B. (2014). *BlandAltmanLeh: Plots (Slightly Extended) Bland-Altman plots.* R package.

Lemon, J. (2006). Plotrix: A package in the red light district of R. *R-News*, 6(4):8–12.

Lenth, R. V. (2015). Using **lsmeans**. Accessed from `cran.r-project.org/web/packages/lsmeans/vignettes/using-lsmeans.pdf` on April 2, 2015.

Lenth, R. V. and Herve, M. (2015). *lsmeans: Least-Squares Means.* R package.

Leslie, P. H. and Chitty, D. (1951). The estimation of population parameters from data obtained by means of the capture-recapture method. I. The maximum likelihood equations for estimating the death-rate. *Biometrika*, 38:269–292.

Lillis, D. A. (2014). *R Graph Essentials.* Packt Publishing, Birmingham, UK.

Lloret, J., Shulman, G., and Love, R. M. (2014). *Condition and Health Indicators of Exploited Marine Fishes.* Wiley Blackwell, West Sussex, UK.

Loff, J. F., Murta, A., and Kell, L. (2014). *ALKr: Generate Age-Length Keys for Fish Populations.* R package.

Mace, P. M. and Sissenwine, J. P. (1993). How much spawning per recruit is enough? *Canadian Special Publication in Fisheries and Aquatic Sciences,* 120:101–118.

Maceina, M. J. (1997). Simple application of using residuals from catch-curve regressions to assess year-class strength in fish. *Fisheries Research,* 32:115–121.

Maceina, M. J. and Bettoli, P. W. (1998). Variation in Largemouth Bass recruitment in four mainstream impoundments on the Tennessee River. *North American Journal of Fisheries Management,* 18:998–1003.

Maceina, M. J. and Pereira, D. L. (2007). Recruitment. In Guy, C. S. and Brown, M. L., editors, *Analysis and Interpretation of Freshwater Fisheries Data,* chapter 4, pages 121–185. American Fisheries Society, Bethesda, MD.

Maceina, M. J., Rider, S. J., and Lowery, D. R. (1993). Use of a catch-depletion method to estimate population density of age-0 Largemouth Bass in submersed vegetation. *North American Journal of Fisheries Management,* 13:847–851.

Mangel, M. (2006). *The Theoretical Biologist's Toolbox: Quantitative Methods for Ecology and Evolutionary Biology.* Cambridge University Press, Cambridge, UK.

Manly, B. F. J. (1984). Obtaining confidence limits on parameters of the Jolly-Seber model for capture-recapture data. *Biometrics,* 40:749–758.

Marjomaki, T. J. (2004). Analysis of the spawning stock-recruitment relationship of Vendace (*Coregonus albula* (L.)) with evaluation of alternative models, additional variables, biases and errors. *Ecology of Freshwater Fish,* 13:46–60.

Matloff, N. (2011). *The Art of R Programming: A Tour of Statistical Software Design.* No Starch Press, San Francisco, CA.

McBride, R. S. (2015). Diagnosis of paired age agreement: A simulation approach of accuracy and precision effects. *ICES Journal of Marine Science,* 72:2149–2167.

McBride, R. S., Hendricks, M. L., and Olney, J. E. (2005). Testing the validity of Cating's (1953) method for age determination of American Shad using scales. *Fisheries,* 30:10–18.

McBride, R. S., Sutherland, S. J., Merry, S., and Jacobson, L. (2014). Agreement of historical Yellowtail Flounder estimates: 1963-2007. Working Paper 2014/32, Transboundary Resource Assessment Committee.

McDonald, T. (2015). *mra: Analysis of Mark-Recapture Data*. R package.

McNemar, Q. (1947). Note on the sampling error of the difference between correlated proportions or percentages. *Psychometrika*, 12:153–157.

Midway, S. R., Wagner, T., Arnott, S. A., Biondo, P., Martinez-Andrade, F., and Wadsworth, T. F. (2015). Spatial and temporal variability in growth of Southern Flounder (*Paralichthys lethostigma*). *Fisheries Research*, 167:323–332.

Mirai Solutions GmbH (2015). *XLConnect: Excel Connector for R*. R package.

Miranda, L. E. and Bettoli, P. W. (2007). Mortality. In Guy, C. S. and Brown, M. L., editors, *Analysis and Interpretation of Freshwater Fisheries Data*, chapter 6, pages 229–277. American Fisheries Society, Bethesda, MD.

Mittal, H. V. (2011). *R Graphs Cookbook*. Packt Publishing, Birmingham, UK.

Moran, P. A. P. (1951). A mathematical theory of animal trapping. *Biometrika*, 38:307–311.

Morton, R. and Bravington, M. (2008). Comparison of methods for estimating age composition with application to Southern Bluefin Tuna (*Thunnus maccoyii*). *Fisheries Research*, 93:22–28.

Motulsky, H. J. and Ransnas, L. A. (1987). Fitting curves to data using nonlinear regression: A practical and nonmathematical review. *The Federation of American Societies for Experimental Biology Journal*, 1:365–374.

Muir, A. M., Ebener, M. P., He, J. X., and Johnson, J. E. (2008). A comparison of the scale and otolith methods of age estimation for Lake Whitefish in Lake Huron. *North American Journal of Fisheries Management*, 28:625–635.

Mukhin, D., James, D. A., and Luciani, J. (2014). *ROracle: OCI Based Oracle Database Interface for R*.

Murphy, B. R., Brown, M. L., and Springer, T. A. (1990). Evaluation of the relative weight (Wr) index, with new applications to Walleye. *North American Journal of Fisheries Management*, 10:85–97.

Murrell, P. (2011). *R Graphics*. Chapman & Hall/CRC, Boca Raton, FL.

Nash, J. C. (2014). *nlmrt: Functions for Nonlinear Least Squares Solutions*. R package.

Needle, C. L. (2002). Recruitment models: Diagnosis and prognosis. *Reviews in Fish Biology and Fisheries*, 11:95–111.

Nelson, G. A. (2015). *fishmethods: Fishery Science Methods and Models in R*. R package.

Neumann, R. M. and Allen, M. S. (2007). Size structure. In Guy, C. S. and Brown, M. L., editors, *Analysis and Interpretation of Freshwater Fisheries Data*, chapter 9, pages 375–421. American Fisheries Society, Bethesda, MD.

Neumann, R. M., Guy, C. S., and Willis, D. W. (2012). Length, weight, and associated indices. In Zale, A. V., Parrish, D. L., and Sutton, T. M., editors, *Fisheries Techniques,* Third Edition, chapter 14, pages 637–676. American Fisheries Society, Bethesda, MD.

Newman, R. M. and Martin, F. B. (1983). Estimation of fish production rates and associated variances. *Canadian Journal of Fisheries and Aquatic Sciences*, 40:1729–1736.

Ogle, D. H. and Winfield, I. J. (2009). Ruffe length-weight relationships with a proposed standard weight equation. *North American Journal of Fisheries Management*, 29:850–858.

Olsen, E., French, R., and Ritchey, A. (1995). Hood River and Pelton Ladder evaluation studies. (BPA Report DOE/BP-81758-1), Bonneville Power Administration, Portland, OR.

Ooms, J., James, D., DebRoy, S., Wickham, H., and Horner, J. (2015). *RMySQL: Database Interface and MySQL Driver for R*. R package.

Otis, D. L., Burnham, K. P., White, G. C., and Anderson, D. R. (1978). Statistcal inference from capture data on closed animal populations. *Wildlife Monographs*, 62:1–135.

Parker, B. R., Schindler, D. W., Wilhelm, F. M., and Donald, D. B. (2007). Bull Trout population responses to reductions in angler effort and retention limits. *North American Journal of Fisheries Management*, 27:848–859.

Patterson, K. R. (1992). An improved method for studying the condition of fish, with an example using Pacific Sardine, *Sardinops sagax* (Jenyns). *Journal of Fish Biology*, 40:821–831.

Pauly, D. (1980). On the interrelationships between natural mortality, growth parameters, and mean environmental temperature in 175 fish stocks. *Journal du Conseil International pour l'Exploration de la Mer*, 39:175–192.

Pauly, D. and Gaschutz, G. (1979). A simple method for fitting oscillating length growth data, with a program for pocket calculators. *ICES. CM. Demersal Fish Comittee*, 1979/6:24:1–26.

Pauly, D., Soriano-Bartz, M., Moreau, J., and Jarre-Teichmann, A. (1992). A new model accounting for seasonal cessation of growth in fishes. *Australian Journal of Marine and Freshwater Research*, 43:1151–1156.

Peig, J. and Green, A. J. (2010). The paradigm of body condition: A critical reappraisal of current methods based on mass and length. *Functional Ecology*, 24:1323–1332.

Peng, R. D., Dominici, F., and Zeger, S. L. (2006). Reproducible epidemiologic research. *American Journal of Epidemiology*, 163:783–789.

Pilling, G. M., Kirkwood, G. P., and Walker, S. G. (2002). An improved method for estimating individual growth variability in fish, and the correlation between von Bertalanffy growth parameters. *Canadian Journal of Fisheries and Aquatic Sciences*, 59:424–432.

Pine, W. E., Hightower, J. E., Coggins, L. G., Lauretta, M. V., and Pollock, K. H. (2012). Design and analysis of tagging studies. In Zale, A. V., Parrish, D. L., and Sutton, T. M., editors, *Fisheries Techniques,* Third Edition, chapter 11, pages 521–572. American Fisheries Society, Bethesda, MD.

Pine, W. E., Pollock, K. H., Hightower, J. E., Kwak, T. J., and Rice, J. A. (2003). A review of tagging methods for estimating fish population size and components of mortality. *Fisheries*, 28:10–23.

Pitcher, T. J. and MacDonald, P. D. M. (1973). Two models for seasonal growth in fishes. *Journal of Applied Ecology*, 10:599–606.

Pollock, K. H., Nichols, J. D., Brownie, C., and Hines, J. E. (1990). Statistical inference for capture-recapture experiments. *Wildlife Monographs*, 107:1–97.

Pope, K. L. and Kruse, C. G. (2007). Condition. In Guy, C. S. and Brown, M. L., editors, *Analysis and Interpretation of Freshwater Fisheries Data*, chapter 10, pages 423–471. American Fisheries Society, Bethesda, MD.

Pratt, T. C., Gardner, W. M., Pearce, J., Greenwood, S., and Chong, S. C. (2014). Identification of a robust Lake Sturgeon (*Acipenser fulvescens* Rafinesque, 1917) population in Goulais Bay, Lake Superior. *Journal of Applied Ichthyology*, 30:1328–1334.

Quinlan, H. (1999.). Biological characteristics of coaster Brook Trout at Isle Royale National Park, Michigan, 1996-98. U.S. Fish and Wildlife Service, Ashland Fishery Resources Office, Ashland, WI, Office Report.

Quinn, G. P. and Keough, M. J. (2002). *Experimental Design and Data Analysis for Biologists*. Cambridge University Press, Cambridge, UK.

Quinn II, T. J. and Deriso, R. B. (1999). *Quantitative Fish Dynamics*. Oxford University Press, Oxford, UK.

Quist, M. C., Guy, C. S., Pegg, M. A., Braaten, P. J., Pierce, C. L., and Tavnichek, V. T. (2002). Potential influence of harvest of Shovelnose Sturgeon populations in the Missouri River system. *North American Journal of Fisheries Management*, 22:537–549.

Quist, M. C., Pegg, M. A., and DeVries, D. R. (2012). Age and growth. In Zale, A. V., Parrish, D. L., and Sutton, T. M., editors, *Fisheries Techniques*, Third Edition, chapter 15, pages 677–731. American Fisheries Society, Bethesda, MD.

R Development Core Team (2015a). *foreign: Read Data Stored by Minitab, S, SAS, SPSS, Stata, Systat, Weka, dBase*. R package.

R Development Core Team (2015b). R: A Language and Environment for Statistical Computing, v3.2.0. R Foundation for Statistical Computing, Vienna, Austria.

Ranney, S. H., Fincel, M. J., Wuellner, M. R., VanDeHey, J. A., and Brown, M. L. (2010). Assessing length-related bias and the need for data standardization in the development of standard weight equations. *North American Journal of Fisheries Management*, 30:655–664.

Ranney, S. H., Fincel, M. J., Wuellner, M. R., VanDeHey, J. A., and Brown, M. L. (2011). Assessing length-related biases in standard weight equations: Response to comment. *North American Journal of Fisheries Management*, 31:661–665.

Rennie, M. D. and Verdon, R. (2008). Development and evaluation of condition indices for the Lake Whitefish. *North American Journal of Fisheries Management*, 28:1270–1293.

Richards, F. J. (1959). A flexible growth function for empirical use. *Journal of Experimental Botany*, 10:290–300.

Ricker, W. E. (1954). Stock and recruitment. *Journal of the Fisheries Research Board of Canada*, 11:559–623.

Ricker, W. E. (1973). Linear regression in fisheries research. *Journal of the Fisheries Research Board of Canada*, 30:409–434.

Ricker, W. E. (1975). Computation and interpretation of biological statistics of fish populations. Technical Report Bulletin 191, Bulletin of the Fisheries Research Board of Canada.

Ripley, B. and Lapsley, M. (2015). *RODBC: ODBC Database Access*. R package.

Ritz, C. and Streibig, J. C. (2008). *Nonlinear Regression with R*. Springer, New York, NY.

Rivest, L.-P. and Baillargeon, S. (2014). *Rcapture: Loglinear Models for Capture-Recapture Experiments*. R package.

Robson, D. S. and Chapman, D. G. (1961). Catch curves and mortality rates. *Transactions of the American Fisheries Society*, 90:181–189.

Robson, D. S. and Regier, H. A. (1968). Estimation of population number and mortality rates. IBP Handbook No. 3, pages 124–158. Blackwell Scientific Publications, Oxford.

Roff, D. A. (1980). A motion for the retirement of the von Bertalanffy function. *Canadian Journal of Fisheries and Aquatics Sciences*, 37:127–129.

RStudio (2015a). Data wrangling cheatsheet. Accessed from http://www.rstudio.com/resources/cheatsheets/ on June 18, 2015. Online Resource.

RStudio (2015b). RStudio: Integrated development environment for R. Accessed from http://www.rstudio.org/ on May 2, 2015. Computer software.

Sarkar, D. (2008). *Lattice: Multivariate Data Visualization with R*. Springer, New York, NY.

Schaalje, G. B., Shaw, J. L., and Belk, M. C. (2002). Using nonlinear hierarchical models for analyzing annulus-based size-at-age data. *Canadian Journal of Fisheries and Aquatic Sciences*, 59:1524–1532.

Schaeffer, J. S. (2004). Population dynamics of Bloaters *Coregonus hoyi* in Lake Huron, 1980–1998. *Annales Zoologici Fennici*, 41:271–279.

Schnabel, Z. E. (1938). The estimation of the total fish population of a lake. *American Mathematician Monthly*, 45:348–352.

Schnute, J. (1983). A new approach to estimating populations by the removal method. *Canadian Journal of Fisheries and Aquatic Sciences*, 40:2153–2169.

Schnute, J. T. (1981). A versatile growth model with statistically stable parameters. *Canadian Journal of Fisheries and Aquatic Sciences*, 38:1128–1140.

Schnute, J. T. and Richards, L. J. (1990). A unified approach to the analysis of fish growth, maturity, and survivorship data. *Canadian Journal of Fisheries and Aquatic Sciences*, 47:24–40.

Schofield, M. R., Barker, R. J., and Taylor, P. (2013). Modeling individual specific fish length from capture-recapture data using the von Bertalanffy growth curve. *Biometrics*, 69:1012–1021.

Schumacher, F. X. and Eschmeyer, R. W. (1943). The estimation of fish populations in lakes and ponds. *Journal of the Tennessee Academy of Sciences*, 18:228–249.

Seber, G. A. F. (1965). A note on the multiple recapture census. *Biometrika*, 52:249–259.

Seber, G. A. F. (2002). *The Estimation of Animal Abundance and Related Parameters*. The Blackburn Press, Caldwell, NJ, second (reprint) edition.

Sekhon, J. S. (2011). Multivariate and propensity score matching software with automated balance optimization: The Matching package for R. *Journal of Statistical Software*, 42(7):1–52.

Shelton, A. O. and Mangel, M. (2012). Estimating von Bertalanffy parameters with individual and environmental variations in growth. *Journal of Biological Dynamics*, 6(supplement 2):3–30.

Shepherd, J. G. (1982). A versatile new stock-recruitment relationship for fisheries and construction of sustainable yield curves. *Journal du Conseil International pour l'Exploration de la Mar*, 40:67–75.

Slipke, J. W., Matrin, A. D., Pitlo Jr., J., and Maceina, M. J. (2002). Use of the spawning potential ratio in freshwater fisheries management: The upper Mississippi River Channel Catfish fishery. *North American Journal of Fisheries Management*, 22:1295–1300.

Smith, M. W., Then, A. Y., Wor, C., Ralph, G., Pollock, K. H., and Hoenig, J. M. (2012). Recommendations for catch-curve analysis. *North American Journal of Fisheries Management*, 32:956–967.

Sokal, R. R. and Rohlf, F. J. (1995). *Biometry*. W.H. Freeman and Company, New York, third edition.

Somers, I. F. (1988). On a seasonally oscillating growth function. *Fishbyte*, 6(1):8–11.

Spiess, A.-N. (2014). *propagate: Propagation of Uncertainty*. R package.

Sprugel, D. G. (1983). Correcting for bias in log-transformed allometric equations. *Ecology*, 64:209–210.

Stari, T., Preedy, K. F., McKenzie, E., Gurney, W. S. C., Heath, M. R., Kunzlik, P. A., and Speirs, D. C. (2010). Smooth age length keys: Observations and implications for data collection on North Sea Haddock. *Fisheries Research*, 105:2–12.

Subbey, S., Devine, J. A., Schaarschmidt, U., and Nash, R. D. M. (2014). Modelling and forecasting stock-recruitment: Current and future perspectives. *ICES Journal of Marine Science*, 71:2307–2322.

Sutton, S. G., Bult, T. P., and Haedrich, R. L. (2000). Relationships among fat weight, body weight, water weight, and condition factors in wild Atlantic Salmon parr. *Transactions of the American Fisheries Society*, 129:527–538.

Szalai, E. B., Fleischer, G. W., and Bence, J. R. (2003). Modeling time-varying growth using a generalized von Bertalanffy model with application to Bloater (*Coregonus hoyi*) growth dynamics in Lake Michigan. *Canadian Journal of Fisheries and Aquatic Sciences*, 60:55–66.

Terceiro, M. and Ross, J. L. (1993). A comparison of alternative methods for the estimation of age from length data for Atlantic Coast Bluefish (*Pomatomus saltatrix*). *Fishery Bulletin*, 91:534–549.

Then, A. Y., Hoenig, J. M., Hall, N. G., and Hewitt, D. A. (2015). Evaluating the predictive performance of empirical estimators of natural mortality rate using information on over 200 fish species. *ICES Journal of Marine Science*, 72:82–92.

Troynikov, V. S., Day, R. W., and Leorke, A. M. (1998). Estimation of seasonal growth parameters using a stochastic Gompertz model for tagging data. *Journal of Shellfish Research*, 17:833–838.

Venables, W. N. and Ripley, B. D. (2002). *Modern Applied Statistics with S*. Springer, New York, fourth edition.

Verzani, J. (2014). *Using R for Introductory Statistics*. Chapman & Hall/CRC, Boca Raton, FL, second edition.

Vigliola, L. and Meekan, M. G. (2009). The back-calculation of fish growth from otoliths. In Green, B. S., Mapstone, B. D., Carlos, G., and Begg, G. A., editors, *Tropical Fish Otoliths: Information for Assessment, Management, and Ecology*, number 11 in Reviews: Methods and Technologies in Fish Biology and Fisheries, chapter 6, pages 174–211. Springer, New York, NY.

von Bertalanffy, L. (1938). A quantitative theory of organic growth (inquiries on growth laws II). *Human Biology*, 10:181–213.

Walford, L. A. (1946). A new graphic method of describing the growth of animals. *Biological Bulletin*, 90:141–147.

Wang, Y.-G. (1998). An improved Fabens method for estimation of growth parameters in the von Bertalanffy model with individual asymptotes. *Canadian Journal of Fisheries and Aquatic Sciences*, 55:397–400.

Wang, Y.-G., Thomas, M. R., and Somers, I. F. (1995). A maximum likelihood approach for estimating growth from tag-recapture data. *Canadian Journal of Fisheries and Aquatic Sciences*, 52:252–259.

Warnes, G. R., Bolker, B. M., Gorjanc, G., Grothendieck, G., Korosec, A., Lumley, T., MacQueen, D., Magnusson, A., and Rogers, J. (2015). *gdata: Various R Programming Tools for Data Manipulation*. R package.

Wege, G. W. and Anderson, R. O. (1978). Relative weight (Wr): A new index of condition for Largemouth Bass. In Novinger, G. D. and Dillard, J. G., editors, *New Approaches to the Management of Small Impoundments*, volume 5 of *Special Publication*, pages 79–91. American Fisheries Society.

Weisberg, S. (2014). *Applied Linear Regression*. John Wiley & Sons, Inc., Hoboken, NJ, fourth edition.

Westerheim, S. J. and Ricker, W. E. (1978). Bias in using age-length key to estimate age-frequency distributions. *Journal of the Fisheries Research Board of Canada*, 35:184–189.

White, G. C., Anderson, D. R., Burnham, K. P., and Otis, D. L. (1982). Capture-recapture and removal methods for sampling closed populations. Technical report, Los Alamos National Laboratory Rep. LA-8787-NERP, Los Alamos, NM.

White, G. C. and Burnham, K. P. (1999). Program MARK: Survival estimation from populations of marked animals. *Bird Study*, 46 (Supplement):120–138.

Wickham, H. (2009). *ggplot2: Elegant Graphics for Data Analysis*. Springer, New York, NY.

Wickham, H. (2011). The split-apply-combine strategry for data analysis. *Journal of Statistical Software*, 40(1):1–29.

Wickham, H. (2014a). *Advanced R*. Chapman & Hall/CRC, Boca Raton, FL.

Wickham, H. (2014b). Tidy data. *Journal of Statistical Software*, 59(10):1–23.

Wickham, H. (2014c). *tidyr: Easily Tidy Data with spread() and gather() Functions*. R package.

Wickham, H. (2015). *readxl: Read Excel Files*. R package.

Wickham, H. and Francois, R. (2015a). *dplyr: A Grammar of Data Manipulation*. R package.

Wickham, H. and Francois, R. (2015b). *readr: Read Tabular Data*. R package.

Wickham, H., James, D. A., and Falcon, S. (2014). *RSQLite: SQLite Interface for R*. R package.

Wickham, H. and Miller, E. (2015). *haven: Import SPSS, Stata and SAS Files*. R package.

Williams, B. K., Nichols, J. D., and Conroy, M. J. (2002). *Analysis and Management of Animal Populations.* Academic Press, San Francisco, CA.

Willis, D. W., Murphy, B. R., and Guy, C. S. (1993). Stock density indices: Development, use, and limitations. *Reviews in Fisheries Science*, 1:203–222.

Wootton, R. J. (1998). *Ecology of Teleost Fishes.* Kluwer Academic Publishers, Dordrecht, The Netherlands, second edition.

Xie, Y. (2013). *Dynamic Documents with R and knitr.* Chapman & Hall/CRC, Boca Raton, FL.

Yates, D., Moore, D., and McCabe, G. (1999). *The Practice of Statistics.* W.H. Freeman and Company, New York, first edition.

Zehfuss, K. P., Hightower, J. E., and Pollock, K. H. (1999). Abundance of Gulf Sturgeon in the Apalachicola River, Florida. *Transactions of the American Fisheries Society*, 128:130–143.

Zippin, C. (1956). An evaluation of the removal method of estimating animal populations. *Biometrics*, 12:163–169.

Zippin, C. (1958). The removal method of population estimation. *Journal of Wildlife Management*, 22:82–90.

Zuur, A., Ieno, E. N., Walker, N., Saveliev, A. A., and Smith, G. M. (2009). *Mixed Effects Models and Extensions in Ecology with R.* Springer, New York, NY.

Subject Index

R Functions (Demonstrated) Index

R Functions (Mentioned) Index

Scientific Names

American Shad (*Alosa sapidissima*), 76

Bay Anchovy (*Anchoa mitchilli*), 217
Black Crappie (*Pomoxis nigromaculatus*), 24
Black Drum (*Pogonias cromis*), 222, 235
Bloater (*Coregonus hoyi*), 52, 56
Bluegill (*Lepomis macrochirus*), 26, 108
Brook Trout (*Salvelinus fontinalis*), 199, 206
Bull Trout (*Salvelinus confluentis*), 51

Channel Catfish (*Ictalurus punctatus*), 211, 267
Chinook Salmon (*Oncorhynchus tshawytscha*), 253
Coho Salmon (*Oncorhynchus kisutch*), 83
Creek Chub (*Semotilus atromaculatus*), 89
Cutthroat Trout (*Oncorhynchus clarkia*), 184, 213

Eurasian Watermilfoil (*Myriophyllum spicatum*), 194

Gulf Sturgeon (*Acipenser oxyrinchus desotoi*), 171

Lake Sturgeon (*Acipenser fulvescens*), 274
Lake Trout (*Salvelinus namaycush*), 83, 102
Largemouth Bass (*Micropterus salmoides*), 46, 108, 194, 248

Pink Salmon (*Oncorhynchus gorbuscha*), 254
Pumpkinseed (*Lepomis gibbosus*), 27

Rainbow Trout (*Oncorhynchus mykiss*), 198
Ruffe (*Gymnocephalus cernuus*), 132, 154

Shovelnose Sturgeon (*Scaphirhynchus platorynchus*), 267
Smallmouth Bass (*Micropterus dolomieu*), 47

Vendace (*Coregonus albula*), 253

Walleye (*Sander vitreus*), 174, 270, 272

Yellow Perch (*Perca flavescens*), 26, 253